Ungendering Technology

This book offers fresh insight into women's mastery of technologies commonly associated with men, with important implications for institutional efforts to identify and support technical proficiency among girls and women.

The work is structured across five original case studies featuring: breast cancer survivors in Newfoundland who constructed a wooden dragon boat using hand and power tools; Egyptian women who used information and communication technologies for political action during the Revolution of 2011; pioneer female audio engineers in the United States working in live concert and studio venues; U.S. female commercial airline pilots who mastered the complexity of flying large aircraft; and a university-educated woman working in sewer maintenance and repair for the City of Detroit in the 1970s. The case studies capture women's own voices and present a range of historical and geographic locations.

A major contribution of this volume is the multidisciplinary analytical framework used to explain women's motivation to engage with non-traditional technologies, the role of peer and political support in encouraging persistence, and informal as well as formal knowledge and skill acquisition. Above all, it is a story of women's empowerment – individually and collectively.

This is a unique book suitable for undergraduates and graduates in the fields of women's and gender studies; science, technology and society (STS) studies; engineering education; and adult education.

Carol J. Haddad is Professor Emerita, Eastern Michigan University. She served as a Professor in the School of Technology and Professional Services Management for 22 years, and also as a faculty affiliate in the Women's and Gender Studies Department, which she headed on an interim basis. She has published on gender and voice in the virtual classroom, and has presented conference papers on student learning outcomes in the *Women and Technology* and *Green Technology* courses she developed and taught online.

Dr. Haddad is recognized internationally for her scholarship on the management of technological change, which generated funded research projects culminating in her 2002 book *Managing Technological Change: A Strategic Partnership Approach* (Sage Publications). She was elected to the Honor Society of Phi Kappa Phi in 2013. Prior to arriving at EMU in 1993, she was a tenured faculty member in labor studies at Michigan State University, and held senior research management positions in the private sector.

She holds a Ph.D. from the University of Michigan (Ann Arbor) in higher and adult continuing education, and an M.S. in labor studies from the University of Massachusetts at Amherst.

Interdisciplinary Research in Gender

Reflective Reading and the Power of Narrative
Producing the Reader
Karyn Sproles

Actresses and Mental Illness
Histrionic Heroines
Fiona Gregory

Cultural Production and the Politics of Women's Work in American Literature and Film
Polina Kroik

Masculinities and Desire
A Deleuzian Encounter
Marek Wojtaszek

Feminism, Republicanism, Egalitarianism, Environmentalism
Bill of Rights and Gendered Sustainable Initiatives
Yulia Maleta

Ungendering Technology
Women Retooling the Masculine Sphere
Carol J. Haddad

Ageing and Contemporary Female Musicians
Abigail Gardner

https://www.routledge.com/Interdisciplinary-Research-in-Gender/book-series/IRG

Ungendering Technology
Women Retooling the Masculine Sphere

Carol J. Haddad

Routledge
Taylor & Francis Group

LONDON AND NEW YORK

First published 2019 by Routledge

2 Park Square, Milton Park, Abingdon, Oxon, OX14 4RN

605 Third Avenue, New York, NY 10017

Routledge is an imprint of the Taylor & Francis Group, an informa business

First issued in paperback 2020

British Library Cataloguing in Publication Data
A catalogue record for this book is available from the British Library

Library of Congress Cataloging-in-Publication Data
Names: Haddad, Carol Joyce, author.
Title: Ungendering technology : women retooling the masculine sphere / Carol J. Haddad.
Description: First edition. | Abingdon, Oxon : Routledge, 2019. | Includes bibliographical references and index.
Identifiers: LCCN 2019009964 (print) | LCCN 2019013673 (ebook) | ISBN 9780429273384 (master ebook) | ISBN 9780367221287 (hardback : alk. paper) | ISBN 9780429273384 (ebook)
Subjects: LCSH: Women in technology.
Classification: LCC T36 (ebook) | LCC T36 .H33 2019 (print) | DDC 331.4/8--dc23
LC record available at https://lccn.loc.gov/2019009964

ISBN: 978-0-367-22128-7 (hbk)
ISBN: 978-0-367-78570-3 (pbk)

Typeset in Sabon
by Taylor & Francis Books

Contents

Foreword and acknowledgments   vi

Introduction   1

1 Built for life: The story of the Avalon Dragon Boat Builders   15

2 Digital megaphone: Egyptian women's cyberactivism in the
   Revolution and beyond   45

3 Sound sisters: Engineering women's music   79

4 Woman in underground Detroit: The non-traditional early
   occupation of a university graduate   107

5 Reaching for the sky: Women pilots at major commercial airlines   117

6 Toward a women-and-technology paradigm of empowerment   150

References   160
Index   182

Foreword and acknowledgments

This book was inspired by my longstanding scholarly interest in women's relationships with technology, and more specifically by 21 years of teaching a cross-listed university course entitled "Women and Technology" that I designed in 1993 for the department of Women's and Gender Studies and the School of Technology Studies at Eastern Michigan University. That course addressed technology development and use in a variety of domains, and served as an analytical forum for graduate and undergraduate students to examine cross-disciplinary research literature, both contemporary and historic, in an effort to determine whether technology contributed to women's empowerment or rather whether it reinforced gender-based social and economic power differentials. My background in adult education enabled me to design the course so that students were encouraged to incorporate their own experiences with technology into their analysis of assigned readings and peer-led weekly discussions. In 1999 I migrated the course from the physical classroom to the World Wide Web to permit greater access by our non-traditional student population – most of whom had work and/or family responsibilities. Women comprised the majority of the students, but the course attracted men most semesters as well.

Women's interactions with technology had been the subject of significant discourse among "second-wave" feminist scholars and activists, and writings from that period served as readings for the Women and Technology course initially. In that literature, while women were encouraged to harness technology for their own purposes, there was a recognition that technical innovation was unlikely to benefit or liberate women when embedded in discriminatory social structures (Haddad 1987; Wajcman 1991) despite a persistent "myth" associating technology with progress (Leonard 2003).

Noted feminist sociologist Judy Wajcman astutely observed that while there was a tendency among some second-wave feminists "to portray women as victims of technology," subsequent scholarship on information and communications technologies (ICTs) and biotechnologies stressed "women's agency and capacity for empowerment" (Wajcman 2007, 292). Moreover, the experiences of some of the *Women and Technology* students exemplified certain aspects of the empowerment theme – particularly in the realms of ICT use, gaming, and biological reproduction. An empirical study that a colleague from the College of

Education and I conducted affirmed our observations as instructors that student discussions in fully-online courses – at the time a relatively "new technology" – afforded women opportunities for greater voice and deeper learning than transpired in face-to-face classrooms (Anderson and Haddad 2005).

I began research for this book in 2008, examining existing scholarship in women's studies, science, technology, and society (STS) studies, and labor studies – fields in which I had extensive professional experience – for evidence of women's empowered use of technology. Given the paucity of scholarship integrating perspectives from women's studies and technology studies (Fox, Johnson, and Rosser 2006, vii; Wajcman 2004; Faulkner 2001), I decided to conduct primary case-study research demonstrating women's mastery of non-traditional technologies. I wove into those case studies information from additional bodies of literature: adult learning theory, and in one instance, Arab studies.

Many people were instrumental in helping me identify interviewees, most notably Jane Brown – one of the Avalon Dragon boat builders who generously sponsored a reception to introduce me to the other builders and who provided valuable information thereafter; Kaia Scaggs, Terry Grant, and Boden Sandstrom who offered leads about pioneer women audio engineers; Bothaina Kamel and Hassan El Naggar who shared valuable insights about Egyptian politics and Janet Patton who helped me to make contact with other female pilots at major commercial airlines. Academic colleagues offered words of encouragement and suggestions at critical junctures, among them: Dr. Daryl Hafter – a pioneer in the history of women and technology and in the leadership of the Society for the History of Technology; Dr. Joyce Kornbluh – who established the Program on Women and Work at the University of Michigan and mentored many union women and female labor educators, myself included; and Dr. Sarah Huyvaert – Professor Emerita of Adult Education whose work and collaboration I have long valued.

I am deeply indebted to all of the women and men who consented to be interviewed for this project, most of whom are named in the chapters unless they wished to be given pseudonyms, or unless I determined that revealing their identity might put them at risk in Egypt's volatile political climate. They generously gave of their time and personal insights during the interviews, educated me about their technologies and the contexts of use, and reviewed and offered suggestions on draft chapters in which their stories appeared. I am also grateful to Eastern Michigan University for providing me with one-semester research leaves in 2008 and again in 2012 which provided valuable time for data collection and writing. The Women's and Gender Studies department offered me opportunities for campus presentations on two of the case studies which served to validate my work.

Dr. Margot Duley, Professor Emerita of History at Eastern Michigan University and Dean Emerita of the College of Liberal Arts and Sciences at the University of Illinois Springfield served as an intellectual sounding board throughout the process of researching and writing this book, nursed me through two major illnesses, and served as a mentor through her many years of feminist activism and path-breaking historical scholarship. My gratitude to her is immeasurable.

I am very grateful to Alexandra McGregor, Interdisciplinary Gender Studies Editor at Routledge for her expertise and encouragement in shepherding my manuscript through the review process. She maintained communication with me at each step – a sign of a professional and engaged editor. My thanks are also owed to the Routledge Editorial Board, Editorial Assistant Eleanor Catchpole Simmons, Production Editor Christopher Taylor, the contract staff, and to two anonymous reviewers for their helpful suggestions.

Finally, I wish to recognize the courage and grace of those who are not here to see their stories published in this volume: Donna Howell and Norma Andres, each of whom succumbed to cancer much too early. They touched all who knew them, and it is my hope that their stories will continue to inspire others.

Introduction

Book rationale and purpose

The fields of science, technology, engineering, and mathematics – popularly known as STEM – are recognized as being vital to a nation's economy and to the career success of individuals (Noonan 2017; Smalley 2018). U.S. policy-makers and educators have expressed concern over the persistent under-representation of girls and women in engineering, physical science, and computer science programs and professions, even though their numbers have increased in life sciences and somewhat in mathematics, with more pronounced gender gaps for minority women (National Girls Collaborative Project 2018). In Canada, the pipeline in STEM education has been described as "leaky" (Gjersoe 2018), and women are underrepresented among STEM postsecondary graduates in many other countries as well (Catalyst 2018). Various initiatives have been developed to increase the numbers of girls and women in STEM fields (AAUW 2017; PLEN 2018; Million Women Mentors 2018).

Role modeling and mentoring are known to be key enablers of women's pursuit of postsecondary technical education in the field of computer science (Dennehy and Dasgupta 2017; Margolis and Fisher, 2002). The seminal research of feminist psychologist Jacquelynne Eccles and colleagues beginning in the 1980s demonstrated empirically that women's motivation to pursue STEM fields – particularly science and math – is a function of interrelated psychosocial factors including childhood experiences, family influences, and social role and cultural stereotypes combined as an "Expectancy-Value Model of Achievement-Related Choices" (Eccles [Parsons] et al. 1983; Eccles 1987). That model's inclusion of gender roles and stereotypes advanced what is known about influences on women's education and career choices (Eccles 2011). What has been less clear is how women develop technological expertise outside of K-12 and undergraduate academic settings. Formal education, though obviously important, is only one venue in which technical knowledge is required.

A series of research questions gave rise to the present study. What moti-vates women to engage with technologies that are commonly considered to be the province of men? What are the characteristics of those technologies? To what extent do psychosocial factors such as personal interest, economic need, influence of family/teachers/peers, and/or contextual factors like cultural/political climate contribute to that motivation? What are the formal and informal ways women learn to use and become proficient in the use of non-traditional technologies? How do women overcome the limitations of gender-role stereotypes and patriarchal institutions and cultures to use "masculine" technologies with confidence, and what benefits do they derive in so doing? Finally, is women's technology mastery merely an individual process, or is learning supported by non-hierarchical interactions with peers in keeping – consciously or not – with feminist principles? This volume addresses these questions through original, cross-national case study research.

Chapter 1 presents the stories of female breast cancer survivors in New-foundland, Canada who experienced empowerment and healing through their hand-construction of a wooden dragon boat for racing. Chapter 2 describes Egyptian feminists' mastery and use of information and communications technologies for political action and social change during the period of time surrounding the Egyptian Revolution of 2011. Chapter 3 depicts the ways in which a small number of pioneering women in the U.S. learned the trade of live and studio audio engineering and applied their talents to the production of feminist music. Chapter 4 documents the story of one university-educated woman who in 1979 began work in a physical, non-traditional profession – sewer maintenance and repair for the City of Detroit. Although that chapter is based on a single person, it deserves a spot in this volume because it so richly demonstrates informal learning between an Asian-American woman and her African-American co-worker, and illustrates the uneven career paths available to bright, working-class women – especially minority ones – given certain regional and economic factors. Chapter 5 examines the career paths and technology mastery of U.S. female commercial airline pilots – who continue to be a minority in that profession despite women's long history as aviators. Chapter 6 is a synthesis of lessons learned in relation to the research questions posed and the implications for women's empowered use of technology.

These diverse case studies were chosen to illustrate psychosocial and other factors that motivated women to master and use "masculine" technologies for work, pleasure, healing, and to promote societal equality. The terms "masculine technology" and "non-traditional technology" are used interchangeably throughout this volume and refer to those technologies traditionally associated with men. Technology mastery denotes a high level of proficiency that enables successful completion of a project and carryover of requisite skills to future projects; such proficiency may also be recognized by formal certification.

Relevance of adult learning theory

Efforts to promote women's entry into technical fields tend to center on recruiting them into degree programs at colleges and universities. Yet, as has been the case historically and as illustrated in the case studies contained herein, informal learning is a path to technological literacy for many adults.

Informal learning has been defined as "any activity involving the pursuit of understanding, knowledge or skill which occurs outside the curricula of educational institutions. . . . undertaken either individually or collectively, without the presence of an institutionally authorized instructor" (Livingstone 2000, 2 as cited in Hamilton 2006, 126–127). Informal learning may be intentional or incidental, the latter being "a byproduct or unintended outcome of a learning experience" (English 2002, 232). Among the factors thought to contribute to informal learning are life experience, organizational context, action orientation, non-routine conditions, and proactivity, creativity, and critical reflection (Watkins and Marsick 1992, 287).

Informal learning may also take place in structured workplaces. The concept of workplaces as "learning organizations" – popularized in management circles by Peter Senge (1990) – continues to be considered essential to organization and business success (Hess and Arlow 2014) as a means of increasing performance and agility (Sarder 2016). Informal learning in the workplace "may occur as a result of . . . mentoring, coaching and job shadowing" and "by engaging with others or by embarking on some sort of self-initiated study" with much of it "unplanned . . . because it occurs as needed" (Manuti et al. 2015, 5). Peer interaction can "validate and promote" informal workplace learning (Galanis et al. 2016, 597), though it should be noted that in hostile or sexist environments that is not easy to come by.

Whether learning transpires in informal or formal settings, it can have a transformative effect on those who engage in it. Mezirow (2000) defined transformative learning as:

> the emancipatory process of becoming critically aware of how and why the structure of psycho-cultural assumptions has come to constrain the way we see ourselves and our relationships, reconstituting this structure to permit a more inclusive and discriminating integration of experience and acting upon these new understandings.
>
> (6, as cited in Kitchenham 2008, 109).

In transformative learning theory, special attention is paid to the meaning that adult persons apply to the content and process of learning. Mezirow articulated "four ways to learn: by refining or elaborating our meaning schemes, learning new meaning schemes, transforming meaning schemes, and transforming meaning perspectives" (Mezirow 1994, 224). The first involves learners "working with what they already know by expanding on, complementing, and revising their present systems of knowledge, the second refers to the learning of new meaning schemes that are compatible with existing schemes within the learners' meaning perspectives," and meaning transformation occurs when existing meaning schemes are inadequate for solving a problem and "resolution comes through . . . critical self-reflection of the assumptions that supported the meaning scheme" (Kitchenham 2008, 111–112).

Much that has been written about transformative learning focuses on the role that adult educators must play to bring it about (Mezirow 1997; Parra,

Gutiérrez, and Aldana 2015). Yet self-learning and instruction in informal set-tings were primary methods of learning used by most of the women in this study – topics on which there is a paucity of scholarly writing. Their experi-ences demonstrate a clear link between their mastery of non-traditional tech-nologies and the type of transformative learning processes that Mezirow and others have described.

Although Mezirow's original study was on the success of community college programs designed to facilitate women's reentry into postsecondary education, his theory of transformative learning has been criticized for its emphasis on individual agency without acknowledging social contexts steeped in gender inequity and patriarchy (Clark and Wilson 1991). However, he envisioned an empowering outcome, writing that transformative learning would:

> result in learners motivated to take collective social action to change social practices, institutions, or systems. . . . working in concert with like-minded individuals as well as collectively to effect cultural as well as political change in interpersonal relations, families, organizations, communities, or nations.
>
> (Mezirow 1994, 226)

There is a demonstrated connection between transformative learning and empowerment (Hassi and Laursen 2015; Chen 2012). It is known from Freire's (1971) teachings on "critical consciousness" that members of societally oppres-sed groups have the ability to critically analyze their situations and take action rather than being acted upon (Chronister and McWhirter 2003, 423). In fact, the very feminist movement that produced the writing on technology's adverse effects on women also inspired women to engage in personal and political action – even in technological realms. The case studies that follow in Chapters 1–5 demonstrate ways in which women have done just that.

Empowerment theory

To better understand how the case studies illustrate women's empowered use of technology, it is useful to define the terms "empowerment" and "agency." Suchman (2009) observed that the concept of agency stems from Actor Network Theory and refers to "our capacity for action as humans" adding that "the politics of technology" involves "fundamental assumptions about where agency is located, and whose agencies matter" which is of great relevance to feminist studies of technology (p. 3). Empowerment has been defined as "the process of increasing personal, interpersonal, or political power so that individuals, families and communities can take action to improve their situations" (Gutier-rez 1995, 229 cited by Lee 2007, 1098). The works of Paulo Freire (1971) and Albert Bandura (1986) have contributed to our understanding of political empowerment and personal empowerment, respectively. Empowerment occurs when people are equipped "with the requisite knowledge, skills, and resilient self-beliefs of efficacy to alter aspects of their lives over which they can exercise some control" (Ozer and Bandura 1990, 472). The Dragon Boat builders' case study is of special relevance here, though the skilled uses of non-traditional technologies by all of the women in this study were empowering at a personal level, and for the Egyptian women and the audio engineers in particular at a political level as well.

There is some debate about whether agency is a component of empowerment, or rather a separate construct. Drydyk (2013) has taken the latter position:

> To speak of empowerment as a result or outcome presupposes a process of change to produce it . . . with a specific kind of outcome. In this respect it differs already from 'agency', which refers either to a given person's degree of involvement in a course of action. . . . Thus 'agency' refers to a state of affairs while 'empowerment' refers to a process of change.
>
> (251)

Parmar (2004) views the constructs as interrelated, defining empowerment as "a process of discovering one's internal strength, agency, and capacity to effect change in the institutions, behaviours, and ideologies that form the basis of one's experience of systematic oppression and exploitation in daily life" (124). As Ray (2014) has written: "the idea of empowerment entails a strong notion of independent agency: only a robust agency could lead to women's empowerment" (289). I favor the integrative approach, and will use the term "empowerment" as indicative of "agency" throughout this volume.

Feminist thought on women and technology

As stated in the Foreword, feminist scholars have long wrestled with the question of whether technology contributes to women's empowerment or rather whether it reinforces gender-based social and economic power differentials. Based on a review of historic and contemporary literature from women's and gender studies, science, technology, and society (STS) studies, and labor studies, I argue that three key factors distinguish the victim scenario from one of empowerment: control, context, and knowledge. The field of STS has taught us that technology is not an independent variable, but rather is embedded in social and economic structures, and that social beliefs and values influence decisions about technology development and use (Pacey 1983; Pinch and Bijker 1987; Webster 1991; Pool 1997). This is known as the social construction of technology, or social constructivism, and the aforementioned bodies of literature address this theory from intersecting perspectives.

Labor studies conveys that organizational factors and managerial decisions pertaining to technology design, selection, and implementation adversely affect employees and performance when introduced in top-down ways (Haddad 1989; Thomas 1994; Liker, Haddad, and Karlin 1999; Haddad 2002; Hannon 2013).

Feminist labor historians have documented that as technology enabled the movement of production from workshop and household to factory, women experienced more pronounced gender-based divisions of labor and unequal pay even when labor force attachment and job tasks were comparable (Abbott 1910, 174; Wertheimer 1977; Kessler-Harris 1982; Baxandall, Gordon, and Reverby 1976).

To further address the gender-labor-technology intersection, feminist STS-labor historians active in the Society for the History of Technology (SHOT) formed a special interest group called Women in Technological History (WITH) in 1976 thanks to efforts by SHOT's then-president Daryl Hafter among others (McGaw 1982; Trescott 1979, 19–20). WITH remains an active interest group in SHOT, and members interested in gender may also be affiliated with a newer interest group: Exploring Diversity in Technology's History (EDITH), which provides support to "scholars and scholarship currently underrepresented in the history of technology and SHOT" and focuses on "the fields of race, ethnicity, gender, sexuality, class, and disability . . . and . . . intersectionality" (Society for the History of Technology 2018).

The adverse effects of technology introduction and industrialization on women's economic status were also evident in countries affected by Western development agencies and transnational companies – especially in the context of colonial rule, past or present. Boserup (1970) in her landmark book described how in the 1920s, Ugandan women were disenfranchised of their role as the traditional cultivators of cotton by a European director of agriculture, and in other areas where women continued to farm, "the Europeans neglected to instruct the female cultivators when they introduced new agricultural methods, teaching only the men in an agricultural setting of traditional female farming, with particularly unfortunate results in regions with large male emigration" (54–55). Chaney and Schmink (1976) found similar trends in their study of Latin-American women, observing that "It is widely recognized that the 'technological imperative' carries with it a set of sex-role prescriptions. . . . Women are considered incapable of handling or understanding complex machinery, and thus often are deprived of access to useful tools" (175).

Gender bias is certainly a constructivist influence on technology design and use. Gender stereotypes are culturally determined and set parameters on which technologies are considered appropriate for men versus women. Lohan (2000) and Johnson (2006, 3) referred to this respectively as the "co-construction" or "co-creation thesis" and feminist STS scholar Judy Wajcman (2007) used the term "feminist technoscience" to explain that "people and artefacts co-evolve: the materiality of technology affords or inhibits the doing of particular gender power relations" (295). Although feminist scholars have contested the masculine versus feminine technology binary (Oldenziel 1999; Lubar 1998; Stanley 1993), it remains the case that certain technologies are identified more with men than with women. Moreover, greater societal value has traditionally been ascribed to the skilled use of technologies in the production (male) sphere than those in the domestic (female) sphere, based on "social and economic and cultural power—a power defined by boundaries and exclusions" (Lerman 2010, 899).

The call for women to take control of technology and shape it in accordance with feminist values was visible in feminist writings and presentations at least as early as the 1970s. One example was a conference entitled: *Women and Technology: Deciding What's Appropriate* held in Missoula, Montana in April, 1979 (Smith 1979). Judy Smith, an organizer of that conference, called for feminist values to be included in the "appropriate technology" movement inspired by economist E.F. Schumacher, suggesting that "a technology that saves women's time or expands their role options is almost by definition appropriate, unless it can be shown to be hazardous, or energy consuming" (Smith 1983, 67–68). Smith and feminist scholar Ellen Balka later proposed the development of a "Sex Role Impact Statement (SRIS)" to ensure that technology served to empower rather than restrict women (Smith and Balka 1988, 83–84). Another writer of that era, Corlann Gee Bush (1983), proposed "equity analysis" as a means of measuring technology's worth by virtue of its function, its impact on users, and its economic and social system effects – including on sex roles.

Feminist technology assessment was advocated in Europe as well. Instructors at Britain's Open University developed and published a course for adult women with the following objectives: 1) "to demonstrate that technology is not value-free or genderless" given its societal and historical context; 2) to train women in technology assessment methodology based on feminist assessment criteria; and 3) to "demystify technology, in the sense of laying open its main principles, modes of operation and systems of social control" (Bruce, Kirkup, and Thomas 1984, 47). Danish academic Janine Morgall, with expertise in global health care technology assessment, proposed ways of evaluating whether the technology under consideration or development would have a "potentially liberating or controlling" impact on women's status, and added the need to identify alternatives though "public participation by political movements and public interest groups" (Morgall 1993, 200, 202–203). The concept of "human-centered systems design" – popular in Europe (Brödner 1982; Rosenbrock 1989, 1990; Ehn 1989a and 1989b; Corbett 1990) – was redefined according to feminist values by Green, Owen, and Pain (1993) who argued that workplace technologies should be designed to meet the ergonomic and other needs of female users, and should foster values of social utility and egalitarianism.

In Judy Wajcman's (1991) influential work, *Feminism Confronts Technology*, she too identified the link between cultural context and technology development, specifically the ways in which "technological objects may be shaped by the operation of gender interests" (23). She observed that "the stability and form of artifacts depends on the capacity and resources that the salient social groups can mobilize in the course of the development process" (23). Warning that women are a diverse population and that there can be no unitary conception of feminist values, she called for the "need to go beyond masculinity and femininity to construct technology according to a completely different set of socially desirable values" but adding that "In so far as technology currently reflects a man's world, the struggle to transform it demands a transformation of gender relations" (166).

Demands for technology appropriate to women's needs were not limited to Western nations. Women in the Philippines organized a forum at the "Third Technology for the People Fair" in November 1983 in Manila where they met with product designers and manufacturers on end-user needs, and saw demonstrations of domestic technologies invented by Filipino women inventors (Ceniza 1985, 15). On a larger scale, in the late 1990s the Commission on the Advancement of Women, a US-based alliance of global non-governmental organizations involved in economic development and humanitarian work, developed a "gender audit" assessment methodology to promote gender integration and equity in such activities (Morris, Kindervatter, and Woods 1999).

A natural outgrowth of feminist technology assessment was the notion of women being involved in technology design – either as designers themselves or as consultants based on their experience as first-hand users. This can be seen in the Philippine example presented, and in U.S. initiatives to recruit more young women to study engineering, computer science, and other STEM subjects, as mentioned at the beginning of this chapter. The early, non-adjustable shoulder harnesses in cars were arguably designed with male rather than female passengers and drivers in mind, for industrial automotive engineering was largely a male domain when shoulder belts began to appear in passenger cars in the late 1960s (Stein 2005; Riley et al. 2009; Bix 2013).

Faulkner (2001) eloquently described the "dualistic" nature of technical and engineering work, with "an objectivist rationality associated with emotional detachment and with abstract theoretical (especially mathematical) and reductionist approaches to problem solving" on the masculine side, and "a more subjective rationality associated with emotional connectedness and with concrete, empirical and holistic approaches to problem solving" on the feminine side. Yet she went on to argue that: "both sides of the concrete-abstract dualism are required within engineering and computing practice" (85). She concluded that ". . . we need to develop strategies to intervene in the process of designing new technologies as well as in the context of use" (91), adding that women designers "should be more likely to 'see' the needs of particular female users" (1892).

Still, the presence of women in STEM professions like engineering is not enough to result in technologies that embody feminist values. An all-female team of engineers designed a "Your Concept Car" (YCC) for Volvo in 2004, adding touches to appeal to women such as a place to store a purse, and a few features thought to be attractive to men (Rechtin 2004). But Styhre, Backman, and Börjesson (2005) contended that in an overwhelmingly "masculine and male-dominated" industry such as automotive, projects like the YCC may "on the surface . . . develop the organization towards gender equality but . . . in its latent functions reinstate and reproduce gendered organizational practices" (103–104). Schwartzman and Decker (2008) also analyzed the YCC experiment from a feminist perspective, asserting that:

the new design has taken the exit ramp from the production line, entering the non-industrial realm of the aesthetic. The novel features that receive most attention are precisely those that accommodate the most restrictive roles associated with womanhood. . . . The new automobile, far from challenging paternalism, incorporates design features that reinforce paternalistic expectations.

(113–114)

Women's experiences with information and communications technologies (ICTs) are also shaped by control, context, and knowledge. Early feminist literature on that topic featured debates about whether cyberspace served as a male-dominated sphere with the potential for misogynist cyber violence, or whether it was an empowering place for women (Spender 1995; Wakeford 1997; Scott, Semmens, and Willoughby 2001). On the latter perspective, some second-wave feminist activists saw the potential of information technology to serve as a platform for organizing. In 1981 a group of San Francisco women with technical knowledge who operated under the name *Computerus* established the *National Women's Mailing List* (NWML) because they believed that "just as the Moral Majority has been able to mobilize millions of their supporters using computerized lists, feminists too should use this technology to our advantage" (Lippitt 1981). They encouraged individuals (women and men) and organizations to register by mailing a completed form specifying their interests and listing optional demographic information about themselves (NWML n.d.). A similar project during the 1980s was the creation of TARTS (Teaching Artists to Reach Technological Savvy) – a US-based "national database on women artists [and] a computer bulletin board listing for jobs and exhibition" (Smith and Balka 1988, 90).

By the 1990s, the writings of Donna Haraway (1991) and Sadie Plant (1995) advanced the thesis that women could in fact exert control over and benefit from digital technology, inspiring the concept of cyberfeminism, which refers to women's use of ICTs to meet their needs and bypass traditional power relations (Paasonen 2005; Jones 2002; Rosser 2005). In 1995, the U.S. feminist periodical *Off Our Backs* published a list of cyberfeminist websites in categories such as: "Women Helping Women on the 'Net'" and "Feminist Action", "Women's Studies" (Elliott 1995, 6). An interesting recent phenomenon has been cyberfeminism's potential to bridge generational experiences of second- and third-wave and millennial feminists. Everett (2004) expressed a conviction:

> that . . . exchanges among cyberfeminists, cybergrrrls, and nonwired women are extremely exciting and beneficial. . . . even as older feminists tell younger feminists how to do feminist history and philosophy, younger feminists can tell older feminists how to do cyberfeminist art, 'hactivism,' and technological wizardry.

(1281)

Scholars of that period also believed that the Internet could enable egalitarian cross-national communication. Bosah Ebo (1998) saw its potential to deemphasize "hierarchical political associations, degrading gender roles and ethnic designations, and rigid categories of class communities" and linking people who are socially and geographically isolated, though he also cautioned about its potential to create "cyberghettos" replicating pre-existing hierarchical relationships based on socio-economic class (3). Mexican cultural anthropologist Lourdes Arizpe (1999) encouraged women to understand "cyberculture" and use the Internet "to find new ties . . . to communities across countries and continents" and to ensure that information technology is used to enhance "human well-being rather than strengthening existing power monopolies" (xiii–xv).

There is greater access to the Internet globally than ever before, but a digital divide has persisted; 45.3 per cent of households in developing countries had access in 2018 compared to 80.9 per cent of households in developed nations (International Telecommunication Union 2018). There is also a widening digital gender gap, with access hampered by affordability, inadequate education and skills, and inherent gender biases and socio-cultural norms (OECD 2018). Internet access can be a source of economic and political power for women. Economically, female artisans have used the Internet to market their goods globally (Davis 2007; PEOPLink 2012, 2016; de Madres 2012, 2016). Politically, women in a variety of developing nations have derived agency from their involvement with cyberfeminism and cyberactivism, just as the Egyptian women described in Chapter 2 of this volume did during the 2011 Revolution. I will mention just a few examples.

In Kenya, women's cyberactivism flourished during the violence that followed a disputed 2007 presidential election, with blogs serving as vital sources of information to those inside and outside of that country (Goldstein and Rotich 2008). One notable female blogger, Ory Okolloh, who wrote under the name "kenyanpundit", would also co-found Ushahidi (the Swahili word for *"testimony"*), an open-source, non-profit software platform designed to democratize information flow and "lower the barriers for individuals to share their stories" (Kenyan Pundit 2012; Ushahidi 2008–2016). Another young woman, Juliana Rotich – Ushahidi's other co-founder – used her Twitter channel *Afromusing* to keep people informed about the Kenyan situation (Macha 2008; Rotich 2012).

A spontaneous cyberfeminist *Pink Chaddi Campaign* was sparked by a 2009 incident that occurred in Mangalore, India, in which a group of young women socializing in a pub were physically attacked by men affiliated with Sri Ram Sene – a right-wing orthodox Hindu group serving as morality police intent on upholding traditional Hindu values (Cullum 2012). The leader of the group – Pramod Muthalik – not only praised the attack, but vowed to forcibly marry couples found dating in public on Valentine's Day (Bangalore Bureau 2009). In response, a group of young women formed a Facebook group under a tongue-in-cheek name: the "Consortium of Pub-going, Loose, and Forward Women" (Nisha 2009). Within days the group had attracted over 34,000 members of both genders (Dhawan 2009). The organizers then invited all members to take part offline by sending pink underwear to Muthalik. An organizer of the campaign explained why they chose the word chaddi: "Chaddi is a childish word for underwear and slang for right-wing hardliner. . . . It amused us to embrace the worst slurs, to send pretty packages of intimate garments to men who say they hate us" (Cullum 2012). Nine years after the attack, however, Muthalik and 24 members of his group were acquitted due to "lack of evidence" (Misquith 2018). Still, Sri Ram Sene's efforts to win a significant number of elected seats on the Karnataka Legislative Assembly were largely unsuccessful (Tewari 2018).

Women in the Middle East and North Africa (MENA) have for some time used a variety of ICTs – print and electronic media, television, radio, film, and the Internet – to advance their causes within and across nations (Skalli 2006; Stephan 2013). In 2006, Iranian women made use of blogs and other ICTs to demand equal rights, human rights, democratic reforms, and an end to police brutality, and launched an online campaign to collect one million signatures to end discriminatory laws (Shirazi 2012). The Internet "has been an essential outlet for women's movements, in particular since the journals Zan, in 1999, Zanân in 2008 and Irandokht in 2009 were censored" (Direnberger 2011, 3), and the use of blogs and Facebook by women's rights activists and mothers has been described as "the Achilles heel of the Islamic Republic of Iran" (Gheytanchi 2015, 46). As is true in many countries, Iranian bloggers and journalists have paid a high price for their outspokenness (Gladstone and Afkhami 2012; Parastoomarzieh blogspot 2012; Mufta Editors 2012; Kamali Dehghan 2014; Reporters Without Borders 2018), and political reforms have not resulted in freedom of the press (Committee to Protect Journalists 2016). Still, as Karimi (2018) has written, use of the Internet by the Iranian women's movement has played "a significant role as a variable in mobilization and collective action" (162). As an example, she described the "Campaign to Change the Male Face of Parliament" which sought 50 seats for women in the 2016–2020 Parliament; while not reaching that goal, "it was successful in providing the largest female representation in the history of Iran's [290-seat] parliament" (159). President Hassan Rouhani, re-elected in 2017 with the backing of many women, issued an order requiring the appointment of women to 30 per cent of managerial posts from an existing 17 per cent, but according to a female member of the Parliament, that order was ignored (Radio Farda 2018).

Use of the Internet enabled Saudi Arabian women to extend their communication from a physical private sphere that is culturally gender-segregated to a virtual public sphere – in some cases advocating for social change in the process. In May, 2011, an information security specialist and divorced single mother named Manal al-Sharif launched a social media campaign to demand that Saudi women be allowed to drive. Muslim women who supported the campaign argued that the driving ban has no basis in religious teachings "Since there were no 'cars' 1400 years ago. . . . and Women used horses and camels to travel during the time of Prophet Muhammad. . . ." (Zuberi 2011, 1). Ms. al-Sharif subsequently faced arrest and a few days in jail following her act of cultural disobedience; an international campaign calling for her release ensued, and King Abdullah ordered such (Malik 2011; Khalid 2012). There was subsequently a backlash against women driving (Staufenberg 2016), which adversely affected Ms. al-Sharif and others. Although the ban was formally lifted by government decree on June 24, 2018, women's rights activists have been harassed and arrested (Al-Khamri 2018; al-Sharif 2018).

Shalhoub-Kevorkian (2011) examined use of the Internet among young Palestinian college women and mental health workers (76 per cent of whom were women) in occupied Jerusalem. The rich stories she collected documented the varied ways that the Internet has served to empower them – through knowledge acquisition and information sharing, communication with family and friends (especially in times of crisis as when Israeli settlers invade their homes), dealing with Israeli bureaucracies, confidentially reporting abuse or hardship to mental health workers, and as a means to search for employment opportunities. Although the study participants described the dangers of cyberspace in the context of occupation – the ways in which "information posted about their houses, family, or siblings is used against them by the Israeli security apparatus" (195), they revealed that the Internet has also served as "a political tool for organizing their resistance and sharing it with those who are willing to listen" as in the case of publicizing one's own house demolition taped with a mobile phone (197–198). The author concluded that while "the use of technologies eases some conflict-related traumas, facilitates activism, and opens up new venues for e-resistance, it also widens the gap between the haves and the have nots" and reinforces "gendered, raced, and classed hierarchies" thereby leaving open the question of whether empowerment can occur "in a context where time, place, and space are dominated by occupying forces" (201–202).

In Lebanon, the Internet has enabled feminists and sexual minority groups to link with one another and publicly advocate for equality and human rights. "Nasawiya" is, according to its Twitter page, "An organized collective of feminists – based in Beirut – working on gender justice" particularly on causes pertaining to women's health, economic well-being, eradication of sexism, sexual harassment and gender violence, support of domestic migrant workers, and engagement in the political process, among others (Nasawiya 2017a and 2017b).

Helem, the Arabic acronym for "Lebanese Protection for Lesbians, Gays, Bisexuals and Transgenders (LGBT)" is a non-profit organization formed in 2004 to protect LGBT people against "legal, social and cultural discrimination"; its stated goals are to: 1) "raise awareness on HIV/AIDS and STIs among LGBTs and to place the sexual health concerns of sexual minorities on the agendas of policy-makers and health practitioners"; 2) "counter the lack of information (particularly in Arabic) and the pervasive misinformation about homosexuality by providing objective, factual information, initiating dialogue, and refuting common misconceptions about homosexuality"; and 3) seek annulment of Article 534 of the Lebanese Penal Code, which punishes "unnatural sexual intercourse" (Helem 2017). The illegality of homosexuality had prevented LGBT people from organizing openly, though a *gaylebanon* Internet group was established in 1998, and a *club-free* underground social group followed (Azzi 2011). Helem worked closely with the Ministry of Health, and in 2008 that agency became the first governmental institution to advocate that Article 534 be rescinded (Ibid.).

Lesbian, bisexual, transgender, and queer women found the need to form a separate, private support group and in 2007 established "Meem." Its stated goal was "to create a safe space in Lebanon where queer women and transgender persons can meet, talk, discuss issues, share experiences, and work on improving their lives and themselves" (Meem 2010). The need to form a women-only organization was due to the dual oppressions that women face – based on gender and sexual orientation (Zeidan 2010, 150 and 259). Meem's rationale and strategy have been documented by Meem's founder Nadine M (M 2010). Meem continues to have an online presence on Twitter that was last updated in 2015 (Meem 2017), but its earlier website appears to be defunct.

As the cited literature suggests, key determinants of whether technology contributes to women's empowerment or rather whether it reinforces gender-based social and economic power differentials are the ability to exert control over its introduction and use, cultural context, and knowledge. Clearly, women's control over technology has advanced beyond the limitations imposed by earlier periods of industrialization and development. Women in all regions of the world have used the Internet in ways that are empowering. Yet the existing literature on cyberactivism does not drill down to the individual level to identify psycho-social factors influencing motivation to engage with ICTs, ways of learning how to use those technologies, and their limitations. The Egypt case study contained herein is unique in that regard. The other case studies offer similar insights about motivation and learning pertaining to "masculine" technologies used in workplaces and other settings.

Research methodology

The case studies are based on formal interviews averaging an hour in length conducted by this author from 2008–2014 in person, by telephone, and by Skype. Each chapter offers a cross-sectional snapshot of technology use during a fixed period of time. Common questions were asked of all interviewees, with some customization based on technology type and circumstances of use. The common questions addressed the nature of the technology used, how learning and mastery occurred, personal and other life factors (family, formal education, informal learning, and mentoring) that encouraged or discouraged technology use and mastery, and benefits derived from the technological proficiency acquired. Interviews were recorded with the permission of interviewees, and were transcribed personally by this author for reasons of confidentiality and accuracy. Following university human subjects protocol, each interviewee gave written consent to the researcher to use her or his words in the book manuscript, with identities masked when requested or where identification may have put the interviewee at risk. Extensive analysis of literature from women's and gender studies, science, technology, and society (STS) studies, labor studies, and adult education provided a foundation for the interview questions and interpretation of the qualitative data they yielded. The case studies that follow further illustrate the benefits women have derived from the technical knowledge acquired, and the empowerment experienced by their mastery of technology in the masculine sphere.

Unique contributions of this book

The case studies herein illustrate the ways in which women's control over varied technologies in different countries and decades has contributed to their empowerment. One unique feature of this volume is the micro perspective that captures *in women's own voices* the essence of technology mastery and the benefits of its use. A second is the variety of technologies studied and the range of time periods and geographic locations presented. Two of the case studies captured real-time technology use – interviews while the dragon boat building was in progress, and discussions with pilots still flying commercial aircraft. Another case study consisted of interviews with Egyptian activists during the Revolution – 19 months after the events resulting in President Mubarak's resignation, but in the midst of continuing political upheaval. The women audio engineers discussed their historic and continued use of the technologies used in their work. Only the Detroit water department worker was no longer using the technologies at the time of her interview.

A third unique contribution of this volume is the integrative weaving together of disparate disciplinary perspectives in the examination of women's relationship to technology. As stated earlier in this chapter, literatures from STS, women's and gender studies, labor studies, and adult education framed study questions and interpretation of interview findings, with additional examination of historic and contemporary research on Newfoundland, Egypt, U.S. women's music culture, and women's aviation history. Finally, unlike much of the literature on women and technology, the women's stories in this book convey the varied ways in which motivation, proficiency, and persistence occur in informal learning settings as well as in those that are more structured, and the role that those factors play among women working in technical careers.

By examining what has motivated women to become proficient in the use of technologies discussed in these chapters, it is hoped that this book will offer pragmatic advice and encouragement to women wishing to learn and use technologies that have been the purview of men historically, thereby "ungendering" technology.

1 Built for life

The story of the Avalon Dragon Boat Builders

Despite improved screening in higher-income countries (Altobelli et al. 2017), breast cancer remains "the most frequently diagnosed" form of cancer, and "the leading cause of cancer-related death among women worldwide" (Torre et al. 2017, 448). Among Canadian women, it is "the most common type of cancer" found (Rethink Breast Cancer 2018), and in 2017 an estimated 5,000 women died from breast cancer – 14 per day on average (Canadian Cancer Society 2018). Lymphedema is often one of the consequences of breast cancer surgery when lymph glands in the armpit are removed or damaged from radiation. The conventional thinking was that the severe upper arm swelling and discomfort resulting from lymphedema were likely to be worsened by strenuous upper body exercise (Miedema et al. 2008). Some researchers who questioned that belief found that vigorous physical exercise could be therapeutic to breast cancer survivors both physically and psychologically (Harris 2001; 2012; McKenzie and Kalda 2003; Lane, Worsley, and McKenzie 2005; Holmes et al. 2005; Cheema and Gaul 2006; Ahmed et al. 2006).

Two such scholars working at the University of British Columbia – Dr. Donald McKenzie, professor of sports medicine and former one-time competitor with the Canadian Olympic canoe team, and Dr. Susan Harris, then professor of rehabilitation sciences and physiotherapist – challenged the myth in a very real way (Richards 2000, Kessler 2018). One of McKenzie's experiments involved the creation of a dragon boat team of breast cancer survivors – "Abreast in a Boat" in 1996. The team of 25 women – including Harris who was herself a breast cancer survivor – underwent a special six-week exercise program before beginning their training in the boat, and were monitored by a medical team consisting of a sports-medicine physician, a physiotherapist, and a nurse; additionally, some of the women with pre-existing lymphedema were fitted with special compression sleeves (Harris 2008, 3). According to Harris, they competed as the only all-women's team in the Novice Division at the World Championship Dragon Boat Festival in Vancouver in June, 1996, making a "respectable showing" and showing no adverse lymphedema effects (Harris ibid.).

Since then, breast cancer survivors have embraced dragon boat racing with a passion. By 2005, it was reported that 69 teams existed in Canada, and the sport had spread to Australia and New Zealand (29 and 5 teams respectively),

Europe (England, Italy, and Poland – four teams altogether), Asia (Singapore, Malaysia, and China – four teams in all), and the U.S. (20 teams) (Waves of Hope 2005). Dragon boats are typically 40–48 feet in length (Lo 2008), colorfully decorated with a dragon head at the bow and a tail at the stern, and paddled, not rowed, by 18–20 paddlers (24 in the largest boats) assisted by a drummer-caller and steersperson. Dragon boating originated in China over 2,000 years ago, and experienced a modern revival beginning in 1976 with Hong Kong's sponsorship of an International Dragon Boat Festival to promote tourism (International Dragon Boat Federation 2008).

This chapter documents the story of how breast cancer survivors in Newfoundland[1], Canada built their very own wooden dragon boat – a non-traditional task in a culture in which boatbuilding has historically been the province of men.

It is based on interviews with 11 of the boat builders and two of their husbands who occasionally participated in the build in support of their spouses. Interviews were conducted in person and by telephone primarily during the summer of 2008 when the boat was nearly completed, with follow up that fall. This story is about more than the skill that the women demonstrated in building a wooden dragon boat with hand and power tools, but also what the entire process meant to them.

The birth of the Avalon Dragons

In 2006, Newfoundland and Labrador stood alone with Prince Edward Island as the only two Canadian provinces without dragon boat teams comprised of breast cancer survivors.

In that same year, two Newfoundland women – Julie Bettney, a retired municipal government leader and former provincial cabinet minister, and Betty Ann Vater, a businesswoman – had learned about the existence of dragon boating teams of breast cancer survivors in other provinces through their work on the Board of Directors of the Canadian Breast Cancer Foundation's Atlantic Division. Betty Ann called an informal meeting of about a dozen interested people in March 2006, and those present decided to try to organize a dragon boat team of breast cancer survivors in the provincial capital city of St. John's, located on the island of Newfoundland's Avalon Peninsula. Julie offered to lend her organizing skills to the effort, and emerged as the Chair of the fledgling group.

After laying some groundwork during the spring and summer, the group sponsored a public launch of their organization – the Avalon Dragon Boating Association – on October 30, 2006 at the Johnson Geo Centre in St. John's, netting local media coverage. Twenty-two breast cancer survivors – most with no knowledge of dragon boating – participated. Local feminist filmmaker and activist Gerry Rogers, who would later be elected to the Newfoundland and Labrador House of Assembly, worked with Julie and others to help to organize the event, which featured breast cancer survivors wearing personal flotation devices over white shirts sporting an Avalon Dragon Boating logo, and carrying paddles as they entered the room in a drummer-led procession

According to Julie Bettney, the October 30 event ". . . made a big impression in terms of the public profile, and all of a sudden people were talking about this and wanting to get involved with it." A core organizing group emerged, and within a month there were nearly 60 members (Avalon Dragon Boating 2008, 2018). A proposal to the Canadian Breast Cancer Foundation netted $10,000 in start-up funding for their first boat. The group made contact with Bruce Whitelaw, Chair of the Naval Architecture Diploma Program at Memorial University's Marine Institute (Society of Naval Architects and Marine Engineers 2002, 5) for advice on how to purchase a boat. Involvement with the breast cancer survivors was a natural fit for Bruce, as he would later explain: "My father offered much of his professional life to the British Columbia Cancer Institute as a cancer research physician, so when the Avalon Dragons approached me to become involved I jumped at the chance" (Green 2008, 3). During a meeting of the core group, Bruce put forth a suggestion that took them by surprise – that in keeping with the seafaring culture and tradition of boat-building in the province, the women might consider building their own boat instead of purchasing one. Julie recalled that when Bruce proposed that the women build a dragon boat:

> . . . there was dead silence [in the room]; you could see the wheels turning.
> . . . And then something kind of electric happened, and people started to
> get really excited about it – 'of course we could build a boat' . . . and as we
> started to explore it a bit more. We . . . could only find one other team
> anywhere who'd ever done this, and that was the team in Australia.

Bruce created a small model of the wooden boat that the Avalon Dragons would build; his design was for a boat 42" long, 46" wide, 660 pounds in weight, and with enough room for 20 paddlers, a steersperson, and drummer. Ultimately it would be made of Western red cedar, Douglas fir, ash, and Okoume, coated with epoxy and fiberglass (Whitelaw 2008, 27).

The women's decision to build their own boat was significant in light of Newfoundland's history. By way of background, Newfoundland and Labrador had been a semi-independent British colony until its citizens voted by a narrow margin to confederate with Canada, becoming that country's tenth province in 1949 (Historica Canada 2013). The island of Newfoundland is the most populated part of the province known as Newfoundland and Labrador. Its rugged, stunningly beautiful coast was described by Margaret Duley, Newfoundland's first internationally recognized novelist, as "rock-lipped, high-piled and washed by a greedy sea" (Duley 1949, 1). That very same sea provided a fishing industry that was Newfoundland's primary source of wealth for centuries.

Women were vital to the family economic unit in coastal Newfoundland fishing villages, and were also the backbone of many community organizations in rural and urban areas alThe Newfoundland women's suffrage movement had a strong sense of the importance of women's work as historian Margot Duley explained in her pioneering book: "In outports dependent upon the fishing economy there was another theme woven into the overall suffrage message: the importance of women's traditional work inside the family unit, and yet its simultaneous devaluation" (Duley 1993 86). The traditional division of labor relegated women's tasks to land, stopping at the ocean's edge (Murray 1979). Newfoundland's fishermen often built their own boats, but that would not have been the work of women. The association of boatbuilding as a male tradition was epitomized in a popular Newfoundland folk song that is sung to this day:

> I'se the b'y that builds the boat,
> I'se the b'y that sails her;
> I'se the b'y that catches the fish,
> And brings them home to Lizer.
> (Songs of the Newfoundland
> Outports 1965, 64)

The Avalon Dragon boatbuilding commenced in February, 2007. Approximately 15 women participated on a regular basis, many of them attending all of the three-hour sessions held on Monday, Wednesday, and Thursday evenings and Sunday afternoons each week. Another seven–ten came when they could, health and schedules permitting. The boat build took place in the donated industrial arts shop of the Macdonald Drive Junior High School in St. John's. At the same time – in January, 2007 – those who wished to paddle began physical training at the New World Fitness gym.

Profiles of 11 Dragon Boat Builders

I secured interviews with 11 of the women most actively involved in the boat build: Anne Marie Anonsen, Julie Bettney, Ann Brazil, Jane Brown, Marie Elliott, Sylvia Flood, Pat Greene, Donna Howell, June Ouellette, "Janice" (pseudonym), and "Mary" (pseudonym). In the interviews I asked the same structured questions of each woman: what motivated them to get involved with the Dragons and the boat build, the tasks that they performed on the build, the tools that they used and prior experience with such, their confidence in the use of those tools, their personal/family backgrounds, and the benefits of their participation. The two husbands interviewed – Neil Elliott and Don Ouellette – were queried about their own family and community backgrounds and connection to boatbuilding, women's work in families, and their observations about their wives' participation and resulting confidence. The interviews took place in the summer and fall of 2008, in person and by telephone.

The women interviewed ranged in age from about 40–70, with an average age of 55. They came from diverse backgrounds in terms of where they were raised (city, coastal, and small town communities), and marital status, with some married or partnered, and others divorced. Many of the boat builders were mothers and some were grandmothers. Some were still in the paid labor force and others were retired. Three of the women had held somewhat non-traditional jobs – elected official; police officer; former director of an organization promoting women's involvement in technical fields. The other seven had occupations that were typically occupied by women: bookkeeper/tax accountant; nurse; teacher; physiotherapist; social worker; seamstress; housewife.

Motivation to participate in the dragon boat build

Three of the interviewees attended the kick-off event at the Johnson Geo Centre on October 30, 2006 and joined the Avalon Dragon Boating Association, known informally as the Avalon Dragons, that evening. Jane Brown, a retired physiotherapist, learned about the plan to begin dragon boating among breast cancer survivors through her volunteer work at the Canadian Cancer Society and her connection with Gerry Rogers. Jane was familiar with the work of Dr. Susan Harris, stating:

> I knew of the benefits of dragon boating . . . so I did go to the opening ceremony at the GEO Centre on Signal Hill where a drum was beaten and various participants came in with their paddles and t-shirts and I thought – 'I have to be part of this.' So I . . . signed up to join in the fitness preparation at New World Fitness – a local facility that donated their facility for a year because Julie Bettney had negotiated that, and I actually joined the Dragons with the intent that I would not be paddling, [but] felt I could build the boat and support the process [of bringing dragon boating to Newfoundland].

Anne Marie Anonsen learned about the boat build from her friend Gerry Rogers. One of the presenters at Johnson Geo Centre who impressed her was Dr. Kara Laing, a local oncologist who spoke of the benefit of exercise for survivors suffering from lymphedema. Anne Marie recalled: ". . . up until that point [when Laing stated that] cancer decommissions people . . . I had not acknowledged my cancer . . . the impact it had on my life. . . . [Laing's words] gave me permission to say: 'This was a hard thing that happened to me'". Thereafter she went to where the build was taking place "to check it out" and met the other women and ". . . we met this man who was going to teach us how to build a boat." Anne Marie stated that since her prior work had been "promoting non-traditional occupations for women, so I thought just from that perspective it would be an interesting thing for me to do."

Sylvia Flood attended the Geo Centre event as well, having received a phone call about it from a male friend who was on the Avalon Dragon Boating Association's executive committee. Sylvia stated that the exercise coupled with a desire to immerse herself into a major activity following her second retirement convinced her to join, for she thought that dragon boating would be "a nice activity with a great bunch of women. That was the big motivation – to be part of a group again, and the physical aspect of it." A bonus for her and others was the free one-year membership offered by a local gym – New World Fitness – to members of the Avalon Dragon Boating Association, referred to most commonly as the "Avalon Dragons."

Ann Brazil learned about the Dragons "through the media" having read about it in St. John's newspaper, *The Evening Telegram*. She then went to a meeting and got involved with the group in February–March, 2007, motivated by "the knowledge that it was a boat being built here in Newfoundland and the first time ever in Canada – and it was women doing the actual boat building." Although the time commitment would be substantial, Ann stated that "it was stress relief 'cause I had some medical problems with my husband and with my brother being in the hospital at the same time. . . . It was enjoyable just to go and your mind was relieved."

June Ouellette heard about the boat build from her husband Don, who worked with Ann Brazil. Her motivation in joining was "people – meeting other breast cancer survivors and making new friends." Marie Elliott also found the social aspect of building a boat with other breast cancer survivors attractive, even though she described herself as "really shy about joining things and meeting people and getting together in groups" and didn't discuss her cancer with those outside of her family. When she learned about the project through a local television news broadcast a month before the first group meeting, she wondered:

> Will this be for me? [But] something kept driving me to say 'I'm going to do this.' From the very first meeting, I just felt a part of the group. You just felt something there that you connected with these women right away. . . . this group of women just felt really wonderful to be with right from the beginning because it was positive. . . .

"Mary" (pseudonym) stated that initially her primary motivation in joining the Avalon Dragons was the free year membership at the gym. Her focus was on "getting fit, learning to paddle and being part of the paddling team." Donna Howell, a retired nurse whose breast cancer was fairly advanced, did not attend the initial meeting because "I didn't have any confidence that I would be physically fit enough to do that" (paddle the boat). Then she learned about the plan to build their own boat, and so she went to a meeting where a small prototype of the boat was shown and the architect, Bruce Whitelaw, was there "and I came away from the meeting thinking 'OK – so we'll be building the boat, but there will be real boat builders there, and we'll be kind of handing them tools and holding boards, basically like assistants'." But when she started to go to the build and participate in Bruce's sessions on tool operation and tool safety and she realized that they would be the builders, "and then it went from there."

Prior use of tools

The women's prior use of tools varied. Some women, like Anne Marie Anonsen, Jane Brown, Mary, Ann Brazil, and Sylvia Flood had used hand tools like hammers and screwdrivers for home maintenance tasks, or saws or hatchets for cutting trees or firewood, but not heavy power tools beyond lawnmowers. Ann's 11 years as a single mother earlier in her life before her later marriage to Myles had built self-reliance:

> . . . I learned to do a lot of things on my own, so I did know how to use a hammer, I did know how to use a screwdriver, I did know how to use an . . . electric drill, . . . and I knew how to use a plane . . . and hand sander. . . .

Jane's work as a physiotherapist beginning in the late 1960s had included working with rehabilitation teams to design and make rehabilitative equipment for children and adults using wood, leather, and metal. Limited funds and limited access to equipment required a level of creativity in using materials at hand. Jane also recalled that as the only child in her family for seven years, she was encouraged to use basic hand tools by her father who managed an automotive body shop, and whom she described as having had "a facility for mathematics and building." She recalled that one Christmas she received an unsolicited gift of "a tool package [containing] a hammer, saw, screwdriver, and from time to time I'd help my father build little boats or little objects." When she reached adulthood and was about to leave home, her father gave her his own hammer, screwdriver, and one of his saws "in case of emergency, or in case of need."

A few women, like "Janice" (pseudonym), Julie Bettney, and Donna Howell had used hand and power tools extensively in the home, having been taught by their fathers or learned simply by tackling building projects with their husbands. Janice's father had a workshop behind the house, which she often visited in the evenings after doing her school lessons. He encouraged her to follow her interests and help him. Donna had used chop saws and jigsaws prior to the boat build, and had observed her dad use the router, planer, and radial arm saw. She explained:

> I've always liked doing things with my hands. I've always enjoyed minor . . . carpentry. My husband and I kind of have a deal in that he does most of the big things, like he'll tear down the deck and rebuild another one, whereas I will do the small things. I'll put up the pictures, I'll [fix] a doorframe. And most of the time even when we did the big things if he needed an extra hand I was the one that was there with him, so I learned.

Julie also described herself as having been "familiar with a lot of power tools" like table saws and sanders that she'd taught herself to use for home projects. She and her husband had built an octagonal dome-style shed in their backyard by following a plan, with "neither of us really knowing what we were doing." She herself installed diagonal wood paneling in their basement, and had built wooden bookcases. Julie indicated that she's always enjoyed working with her hands and using "technology," adding that her husband described her as a "gadget person."

Although Donna had prior experience with power tools, she expressed a preference for performing certain tasks manually:

> I tend to like hand work more so that the mechanized stuff. . . . My husband has these power things, like using drills now for putting in screws . . . and I'm still old fashioned; I've got my own screwdrivers and what not, and I still tend to do things the old way.

She explained how her father's knowledge of carpentry tools and techniques had influenced her:

> My father had a lot of the older tools . . . whereas my husband had some of the more mechanized tools. I did more with my dad when I was home, and even now . . . if there's anything that requires a bit of handwork, like we were putting up cove molding and I asked my father how to 'cope' the angle of the molding. Now my husband would tend to use the mitre saw, chop it at 45 [degrees] and there you go, whereas if I want it perfect, I'd go out to dad and say 'How do I do this so that it just fits in like a glove?' And he'll take me downstairs and show me with a couple of pieces how to do it. . . . Dad always worked with the older tools but once I left home he probably ended up with more of the mechanized tools than he did before, but he still tends to go back a lot of the time to more of the old-fashioned things.

It was often the case that in the era in which these and the other women had grown up, families had to rely on themselves and their neighbors to build and repair homes and boats out of economic necessity. Mary recalled that in her childhood of Harbour Main which she described as an "outport" – a term that refers to a largely self-reliant coastal community – men would learn from one another by "hanging around the neighbor's yard" and help if that neighbor was building a boat or barn or shed or fish stage, adding "people were so poor they had to do these things themselves." She recalled that there was a "different culture around the bay." If your neighbor was building a house you would "give them a day" as it was called, of free labor to keep the building cost down, and it was reciprocated by neighbors. Women participated too – even if to make curtains. She commented that the practice is still in place in many of the smaller bayside towns.

Donna's father, who had lost his own fisherman father at sea near their south coast community of Hermitage when he was only 12, learned from his uncle and others to do:

> . . . everything around the house that needed to be done. You had to be kind of Jack-of-all-trades at that time when you didn't have a lot of money. You had to know how to do stuff, or you had to find somebody else and go out and watch them and come back and do it yourself. . . . in those days

they were handy with tools simply because they had to be. They couldn't buy things; they had to make them.

Even when families moved up the socio-economic ladder, they often carried forward building traditions learned from earlier generations. Although Donna's father ultimately graduated from college and became a production manager for a company that owned several fish plants, he had precious memories of watching his father build things. Donna mentioned that her father had "an old toolbox [that] belonged to his father with his father's old tools in it, and he treasures that, needless to say. . . ."

The mothers of some of the interviewees also learned to perform certain home maintenance tasks out of necessity when work took their husbands away for weeks or months at a time. Donna recalled that:

> mom had to know how to do things herself. She had a set of tools in the house – all of the basic things . . . if there was something that you needed and that couldn't be put off until dad got home, then she would do it . . .

. . . such as fixing a broken rung on the stairs. Marie Elliott's mother also had to be "the mom and dad" for ten months of the year during her childhood while her father worked at the Saglek, a U.S. early-warning radar station in Labrador.

Widowhood was another source of women's self-reliance, as Jane explained:

> . . . my father's sisters always after their husbands passed away . . . in their middle years . . . [50s] maintained their own homes. Into their senior years they could paint their homes or fix things. They would plant gardens and flowerbeds, and would have meticulous systems for keeping plants upright.

Two of the women interviewed – June Ouellette and Marie Elliott – came to the build with no prior use of carpentry tools. Pat Greene's prior use of tools was also limited. She described herself as having a "little bit" of knowledge of tools and techniques such as using a nail gun when laying tongue-and-groove hardwood flooring and using a hand saw or hatchet to cut firewood, but she had not previously used the tools involved in the boat build. Pat had an interest in her husband's tools, but had not used them: "My husband has a workshop out in the back of our house. We have all those tools, but I never was part of them, and I didn't know the names of them . . . I would have liked to have learned," adding that her husband may not have had the patience to teach her how to use his tools.

Schooling, family roles, and cultural expectations

Although many of the women had progressive fathers and husbands and self-reliant mothers, some of them experienced curricular limitations in their schools. Many schools of the 1950s and early 1960s in smaller communities offered the core-subject curriculum of reading, writing, mathematics, and science, but not industrial arts, although boys learned to use tools and build things from their fathers, uncles, and grandfathers. Sylvia recalled that:

> Unfortunately in Deer Lake, we were just exposed to the basics; the smaller your community was, the . . . [fewer] courses you had at your disposal. . . . We had nothing other than the classroom, and our courses were just the basic courses that you needed to matriculate.

She explained that the honors students and those who did well in math were encouraged to take physics, and the other students took science.

Still, she noticed that in her small community there was a lack:

> of exposure to the different types of careers you could choose . . . basically for the girls, it would be a teacher, a nurse, or what we called back then 'commercial' which was a secretary. And if you didn't do [one of] these three things, you probably ended up staying in the town of Deer Lake and working in the store.

Although she had done the bookkeeping for the furniture store that her father started and in which both of her parents worked, she opted for a career in teaching, adding: "It took me a long time to realize . . . that that was the career that best suited me" and remarked that the limited career choices made it easier for her than has been the situation for young people today who "have such a struggle to figure out what they want to do."

By the time Ann Brazil was growing up on Bell Island, a place she described with great affection and pride, the high school and district vocational schools she attended offered industrial arts classes that were open to girls, and she readily signed up for "anything that had to do with building." The fact that girls and boys in her community attended separate schools in the earlier grades did not deter her from taking on non-traditional tasks: "I guess from that experience I just took on everything as a woman and participated, so now I haven't got any regrets."

Interestingly, being educated at segregated, all-girls schools gave some of the women confidence in their abilities that carried forward into adulthood. Jane, who attended Bishop Spencer College – a school for girls established by the Church of England – until the age of 14, had little interest in nursing or teaching – the primary career paths for many college-bound women. Jane excelled in science and math, and recalled being mentored by accomplished female teachers and peers. Pat Greene attended a Catholic girls school, noting that:

> I was always with girls. . . . boarding school was all girls, Mount St. Vincent [school] was all girls. . . . and then nursing was all girls, so I didn't have any obstacles about not being able to do whatever; I just never gave it a thought.

In some homes, gender-based divisions of labor were evident. Sylvia noted that during her childhood:

> I didn't see my dad do any housework other than a few dishes. . . . despite the fact that my dad started out his first career as being a cook in the lumber camps, and apparently he was a good cook, but once he moved out of that role, he didn't do it anymore.

Janice's home experience was a bit different; while some tasks were gender-segregated, her parents helped each other with their respective work, as did the children: "we were always helping each other . . . out of necessity."

Ann's efficacy was born in part from the "inspiration" she derived from her mother, who ran the accounting side of the family trucking business even after she was stricken with cancer. Ann was only 12 when her mother succumbed to the disease, but there was a lesson in her loss – the realization that "If you wanted to get anywhere in life, you had to take on different tasks on your own." She took over those bookkeeping functions, remarking that: "I guess that's where my confidence came from, as in taking something and doing it, and being appreciated and having accomplishment at the end of the day."

Still, the career options considered by some of the boat builders may have been influenced by cultural expectations, and choices they willingly made about marrying and having children. Pat – the eldest of the builders – married one year after finishing nursing school, and recognized that being a doctor would have interfered with her desire to have a family: "I wasn't thinking of being the doctor; I was satisfied being the nurse. And I enjoyed working as a nurse every minute that I was nursing." Anne Marie and Julie each tried to widen career opportunities for younger women by encouraging them to enter the skilled trades through the former woman's work with advocacy organizations, and the latter as a provincial government cabinet minister responsible for the status of women.

Formal instruction and informal learning

Learning at the dragon boat build sessions took place through Bruce's verbal explanations and hands-on instruction in the room where the boat was being built. Bruce's training technique was explained by a number of the women: he would set up everything for a particular boatbuilding project and demonstrate how to use the tools on a template or practice piece of wood. He also reviewed the use of certain tools and processes if they hadn't been used in a while. Jane observed that Bruce did extensive preparation for each session, arriving with plans based on measurements, drawings, and occasionally a Plexiglas pattern to be laid on the wood. He also organized details prior to each build session, such as the procurement of wood.

Jane added that Bruce would give the women a holistic explanation of "the principles underlying the task at hand" and then set them forth to accomplish their respective tasks, without hovering over their work. That instructional technique assumed and validated the women's knowledge and ability to take a conceptual framework and turn out individual components of what would become the finished product. It also required an instructor who was able and willing to mentor others. Anne Marie explained:

> . . . this man introduced us to different pieces of machinery, different saws, and then he put his saw down on the table and stepped back and waited for all of us to pick it up, and I thought: 'He's such a gentle soul.' He was so not a pushy person. He was so kind.

Sylvia's words also conveyed the appreciation that she and the others felt for Bruce:

> How fortunate we have been to have found such an angel as Bruce, to dedicate a full year and a half of his life. We missed the occasional boat build; Bruce misses none. And he is such a quiet, kind, caring person. His voice never goes hardly beyond a whisper. He is so caring.

The boat builders were not passive receptacles of Bruce's instruction, but rather were intellectually engaged and would ask for explanations of work processes. Donna recounted questions that she and others asked:

> OK, why are we doing this? What is it that we're accomplishing by doing it? What is it that we're aiming towards? And he would explain it and it would make sense . . . Whatever he said, he said for a reason, and he was usually right, and so we never questioned it [his explanation].

Caution and confidence

Safety in the use of equipment was something that Bruce stressed in his instruction. Pat recalled how she felt on the very first night of the build when Bruce showed the women all of the power saws and other machines:

> It was quite intimidating, I must say, because [when he demonstrated] the great big saw that was very dangerous . . . he said: 'you have to do it this way so you won't lose your fingers or lose a limb'. . . . He stood over us when we were doing every part like when you had to saw a board. . . . He watched us and if we did put our hand in the wrong place he would say 'Don't do it that way because . . . it goes right on and you [may] forget to move your hand. . . . It was a learning process – just learning how to use the machines and getting over the fear of them.

Safety was especially important to women suffering from lymphedema, as Jane explained:

In that meeting [the first at MacDonald Junior High] Bruce introduced the various equipment we would use and my major concern, having lymphedema, knowing that I would have a low ability to recover from a cut or a wound or splinter – so many of us would be at risk – I wondered even if I ought to be part of the build . . . because there are clear directions that even working in the garden one ought to wear gloves for protection – not just from cuts, but from bacterial infections in that extremity. [That fear] has never dissipated . . . I'm always cognizant of wearing protective gloves. I'm always cognizant of material I'm working with, for example fiberglass, which could easily cut you. I'm not concerned with my unaffected side, but I'm always careful using tools.

Jane also described a generalized caution as being "part of my life" relating to her years of treating physiotherapy patients suffering from finger amputations and other hand injuries incurred in accidents with power tools.

A couple of the women believed caution in the use of power tools was gender-related. Mary thought that:

. . . because we were all women who didn't have that male 'I can do anything' attitude and had no experience with these things . . . we were all naturally very cautious around these tools and Bruce was very careful to instruct us in safety around these tools; we would wear goggles if that was indicated, but we didn't touch any of these things unless we had been fully instructed and were well versed in safety. Generally if there were any of the more dangerous tools we would use them in groups of two or three so that if one person forgot something . . . there was somebody who remembered the particular safety rule. And Bruce was always there; he might be at the other end of the room, but he was . . . keeping an eye on us. . . . But once we were instructed in safety . . . we were smart enough not to [work in unsafe ways]; if we were worried about anything, we just stopped.

Anne Marie offered a similar observation:

I feel very safe when Bruce is there and Neil [Marie's husband who participated in much of the build] is there and everyone is watching. We're very insecure with this stuff still, and I think it might be a female thing, because the guys even with little to no experience . . . jump right in, and sometimes hurt themselves and lose fingers and arms. The women are very timid around this stuff. We recognize the potential for damage in these pieces of equipment. [At home] I wouldn't go out and get a table saw unless I had someone really experienced working with me, and that's so much trouble I'd just as soon go get my little hand saw.

Mary added that health considerations entered into the women's judgments about which boatbuilding activities to tackle at a given session; for example, those in chemotherapy with suppressed immune systems might choose to opt out of fiberglassing; those with allergies or asthma would refrain from attending planned sessions involving the use of strong epoxies. Those with lymphedema decided whether to work with vibrating machines such as electric sanders and electric planers, and would modify their activities accordingly. "And of course, nobody had to do anything; if you felt uncomfortable, you just absented yourself from that activity." Ann emphasized: ". . . we don't overdo it . . . we let our bodies tell us what we can do and what we can't do." It is noteworthy that a number of the women commented that their fear of ruining a piece of wood from improper use of a power tool was greater than any concern about self-injury. Still, the women had complete trust and faith in Bruce's instruction, and for good reason. Donna stated: "You knew that Bruce was going to teach you the correct way."

Focus on quality

The women clearly valued Bruce's emphasis on accuracy in the performance of tasks. Marie recounted how he taught them to be precise right from the start, beginning with the construction of the box on which the boat would be built:

> He was telling us to be exact and we were really nervous that everything wasn't going to be measured exactly, and we didn't realize until the box was completed that we didn't have to be exact with that box; that was just a box. But I think he was trying to train us to be precise and exact, because if we weren't exact this whole boat would not line up properly. . . . we measured so many times before we cut. And we were so precise.

Anne Marie added: ". . . we would practice on the cuts that weren't terribly important, and then we would be good with the material . . . when [precision] was really needed. . . . It's a good learning technique."

Mary described the measurement tasks in detail; in building the framework for the boat the women learned to:

> . . . carefully sketch out, measure and cut out using a jig saw the framework on which the boat was going to be built. . . . [and] using various little complicated measuring devices how to measure the angles on each of these pieces of skeleton, because everything had to be angled very precisely in order to get the curvature of the boat, and he taught us how to convert the measurements to actual degrees and then how to plane and jig and route in order to get the angles that we needed.

Jane indicated that the women would invite Bruce to perform quality inspections of their completed work to see if it was good enough because "his knowledge runs deep. . . . He has an objective and learned view." She attributed the women's quest for perfection not only to Bruce's instruction, but in part to culture, stating that she had grown up learning that:

. . . anything that is worth doing is worth doing well. I know Bruce has high standards; I suspect many of us do too. . . . It's the joy of being methodical. . . . We wanted a boat that would be straight, we want a boat that looks ok, but actually we don't mind a few errors, because we've actually made errors, but we could fill those up with a bit of wood and glue. . . . But in our [Newfoundland] heritage we would want a boat that is built the way it ought to be built . . .

Ann stated that "Bruce was an inspirational person" who helped them to realize that a misstep "was never a mistake" and could be corrected (for example, by filling an improperly drilled hole), "so we knew we could do it and enjoy doing it and be happy at the end of the night."

Gender also played a role in the care with which the women performed the boatbuilding and fiberglassing tasks. Neil, who participated in much of the build, observed that:

The ladies did a much better job . . . [on] the finishing work (sanding and polishing) than if the men were doing it. They were more particular and they wanted more perfection. And this type of a build is something that I was never involved with before. The boatbuilding that I've been involved with was more of a rough construction. Cosmetics weren't important when you were building a boat for work.

Informal peer learning

The women learned the techniques of tool use not only from Bruce, but also from one another. Sylvia noted that when "the girls became very confident" they would train one another, or the "boys [husbands who knew how to use tools and who attended some of the build sessions] would teach us." Mary commented that:

Donna had used some of these tools before and she's a natural born tea-cher. . . . [so] I'd often attach myself to Donna in the early days if she was doing something because she had a very gentle, soft-spoken way of passing her skill set on to the next person and [was] encouraging, building your confidence. So yes, we all taught each other.

Ann valued the sharing atmosphere of the boat build, noting that everybody had a chance to do everything, "and that was the excitement."

Skill transference from the domestic sphere

Anne Marie recognized that the boat build "really pulled on a lot of skills we already had that we didn't realize we had – a lot of transferable skills going on there." She added:

I'm a weaver, and another is a quilter and another is a sewer, so we've got a good eye, we're perfectionists, we know when it's done well, we know how awful it looks if it's not done well. I think Bruce really appreciated that. We're good with our hands. I'm really good with a pair of scissors. So these things all ended up being important. How to take something from a pattern and turn it into a 3-dimensional object, how to make it fit, how to make one piece fit to the next piece. . . .

Other interviewees also drew analogies between tasks involved in the boat build and work they had performed in the home. For example, Mary stated that when the boat frame was completed, the beams had to be wrapped in cloths that had been dipped in boiling water, and then the women ironed the wood in order to soften it and form it into the curved shape of the boat, adding that the ironing was "fun." June, a skilled seamstress as had been her mother, likened the cutting of fiberglass (used to make a mold of the wooden boat hull) to "cutting a pattern for sewing" and noted that the precise measuring skills required for the boat build were akin to those used in sewing. Mary also noticed how women applied their domestic skills to the boat build:

Some of the women are really good seamstresses and boatbuilding is so similar to that – the measuring and the ability to look at something and see what it's going to be like when it's finished, and the cutting out of the fiberglass, and the fitting of things together. It's amazing how well they were able to transfer their skills in making a dress to a boat!

Donna, whose quilts had been auctioned internationally through an organization called "The Quilt" to raise funds in support of those living with cancer, also drew comparisons between her quilting and boatbuilding work. She recalled a night when they were working on the boat frame and needed to create "some kind of strange angle and she informed Bruce: I've got rulers at home that have got all of these weird angles on them if that's any good," and thereafter brought in one of her "quilting rulers that had 40, 45, and 50 degree angles" on it, which they used that night. She remarked: "It was so funny, because I remember one of the girls saying afterwards: 'It was so funny, Donna, to see you come in with your quilting ruler' to the boatbuilding."

Technology and task preferences

There were some tools and tasks that appealed to the women more than others. Anne Marie noticed that: "As the boat started to form, we all . . . kind of carved out our niches . . . and had our own different kinds of expertise. . . . Jane and Donna Howell were really good at scarfing, whereas other women were best at planing, others at sanding" adding that every woman played an active and strong part in building the boat.

Janice indicated that she especially "liked the power tools," such as the router. Sylvia's favorite tool was the band saw, a tool she had not used prior to the boat build:

> . . . because it had a small blade, and you had to take a piece of wood . . . and you would have to cut that long piece of wood from one end to the other, really straight. I seemed to have a straight eye for that. [Her accuracy with the band saw earned her the nickname "Dead-Eye Dick"]. And that was a challenge to me – that I could put that [wood] in and stay on that line and have it very precise. . . . I loved that precision.

Marie mentioned that the women "learned how to take apart a plane and sharpen it, so we [learned] to take your time . . . and how to actually sharpen it. [Such tasks were something] that I didn't want to bother with before [because she found them] boring." June stated that she "loved using the router and planning" because those power and hand tools gave her a visual sense of accomplishment.

There were clearly psychological aspects to certain task preferences. Sharpening the plane gave Anne Marie a sense of purpose and fulfillment. She stated passionately:

> I am the queen of the plane – THE queen of the plane. Having said that, I also made sure that I taught every other woman I could who would know how to sharpen the blade of the plane [using a coarse and a fine whetstone]. The plane is a beautiful, beautiful instrument. It is a lovely-looking instrument, it has beautiful lines, it's so sweet the way it comes apart and goes back together again, and I can sharpen the plane so that it just whistles through the wood. And I am so proud of that. . . . Sharpening the plane let me be by myself; it [also] let me be in the group with a function that was important to them, which fed my ego – as soon as I'd walk in: 'Oh Anne Marie, we've got all these planes you've got to sharpen, come over here!' so I'd feel good; I'd feel gratified and appreciated.

Marie, too, appreciated group and individual work. She stated that much of her enjoyment of the boat build came from working with others, seeing something completed, and talking about what they were doing, or "just talking about nothing or anything." When she performed solitary tasks like sanding, she became "lost in my own thoughts . . . and it was meditative when I was just sanding away. A lot of the tasks . . . were just meditative – even the staining. You were in the group, but you were still alone." She reported having the same reaction to housework; she doesn't turn the radio on, and finds it "very meditative – you're alone in your own thoughts" for a couple of hours.

Mary derived satisfaction from "scarfing" the joints of 20-foot long boards to create a strong 40-foot length boat:

they had to be cut, and then planed so they fit together perfectly so that there wasn't even a fraction of a millimeter on either side, so you had to plane and plane and plane both sides and then measure and plane and measure and plane until you got the fitting perfectly. . . . that was really a big deal because that meant the boat was going to be stronger and more secure, and that was fun to do.

She went on to describe other tasks she enjoyed or found tedious:

Using the jigsaw was fun because it gave you such a sense of power – Wow . . . all of a sudden this big piece [of wood] falls off onto the ground! Using the router was fun; you could just see a shape forming when you were routing things. I think all of it was fun; I can't think of any part that I really actively disliked. Sanding, sanding – the many hours we spent sanding; we'd just get so tired after hours and hours of sanding, but even that was fun because it was very collegial; we all were there in a row, and would chat and chat and chat because that was a mindless activity.

Jane especially enjoyed using the router, because she "had never done it before. . . . I simply loved it." She described the router as being initially:

. . . quite intimidating. It's a round machine with a central projection, and it has blades like a blender . . . you must keep it flat, you must keep it stable, upright, 90 degrees, and it will cut the wood in rounded form. It was used to produce smooth, rounded edges on the wood to provide a beautiful and fine finish to the edges of the boat that the paddlers might come into contact with, such as the keel, the bulkheads and the gunwales. . . . We also used the router to join the wood together to make long planks [and] to create finger joints which fit together securely and smoothly.

Those uses of the router gave Jane "instant gratification," for the task was completed the first time through, except for minor touch-ups with sandpaper. She likened the process to "making fine furniture . . . such as the edge of a table" – something her father had done.

Donna, with her preference for hand tools, described her enjoyment of planing: "I liked planing. I hadn't done any hand planing in years and years and years, not since I was home with my dad, and even then it was probably only . . . a couple of strokes at a time." Pat enjoyed using all of the tools, stating: "I was game for it all."

Despite the preferences some of the women had for certain tools, a spirit of sharing and caring prevailed at the boat build sessions. Anne Marie observed that:

Maybe because . . . not one of us was a boat builder [or] carpenter, from the very outset . . . all of us were very conscientious about making sure that everybody had a chance to get on all the tools. So it was always about watching out for the other woman to make sure that nobody was being left behind. And that caring . . . started at the very, very beginning . . . [and] has developed us into a really fine team of women.

Family and community support

The support that the women received from their families was vital to their participation in the boat build. It took various forms, including making dinner, maintaining the household, giving up the companionship of wives/ partners up to four times a week for 17 months, and participating in boat-building sessions. Marie's husband Neil, a computer maintenance technician who had learned basic boatbuilding from his fisherman father, volunteered to assist Bruce in training the women in woodworking and tool skills because "Marie was really excited about the project. So we decided that we would go as a team." He attended about 70 per cent of the boatbuilding sessions, and noticed a distinct increase in the women's confidence as the build progressed, adding: "Those ladies continuously compliment each other on the great job that they're doing."

Anne Marie was heartened by "the incredible support we've gotten from our partners and from our community," which she described in some detail:

> The boatbuilding sometimes has taken over our lives. And we're not there to do the laundry and the cooking and the cleaning, and the shopping for groceries. And I've been leaning on Katja and she's had to be pretty toler-ant, and picked up a lot of pieces, and done more than her fair share of managing our household. And that's not fair . . . except that she just is so happy to be able to be part of it. She's also come down to the boat build often to help us out down there.
>
> But I'm hearing about very conventional women saying the same thing about their husbands who don't know how to cook or clean or do anything like that and now they're going off to the grocery store and coming back with their little bag of groceries – and cooking! And the women are saying: 'I'm home only for a minute, dear, just to get a shower and I'm gone'. . . . it's just another side of how healthy it's all been. It's not the men against the women; it's the men supporting their women. The men and the women supporting their women . . . It's just been a wonderful opportunity for the people in our families to say: 'we want to be part of something that's helping you. We've been part of the nasty, cancer part, but now we want to be part of this healing, healthy stuff.'

Her statement about "women supporting their women" was a reference to sis-ters of the boat builders who participated in the build with them: June's sister Ruth, and Donna's sister Kathy, and to her own life partner Katja. Sylvia mentioned that her husband Roger attended some of the build sessions and was very supportive of her involvement with it.

The boatbuilding experience brought family members together even more closely. June's husband Don, a former Navy sailor from Acadian New Brunswick, noticed the deepening of the sisters' relationship with one another and also benefits to his marriage. He admitted that although at times he tired of hearing about the boat build over those 17 months, he felt that their life together had changed for the better insofar as he "appreciate[d] her more when she [came] home" and they had developed greater mutual "understanding." Women in turn saw their husbands in a new light. Marie stated that prior to the dragon boat build:

> [I] wouldn't let him show me things because I wasn't interested, but when I watch him with the women now, teaching them, and they see him come in and [say] 'Oh great – Neil is here!', I'd never thought of him in that way before. He was just part of me; we'd do things together, but I was so excited to hear the women interested because Neil was there and could show them things. . . . I had never . . . [been] patient enough to let him show me things.

Children living at home supported mothers who had battled cancer prior to and during the boat build. Janice, who had joined the boat build several months into the project, stated that when she lost her hair from chemotherapy treatments, her son – then a high school senior – shaved his head as a gesture of solidarity with his mother.

Community support came in the form of in-kind donations, such as use of the junior high school facility at which the build took place – referred to by Ann as "our shipyard," and the one-year complementary membership at a local fitness gym. Bruce arranged for the donation of wood for the build, including exotic woods from a local business called "Classic Woodwork." Anne Marie described having "access to enough materials" as a "glorious" new experience for her, since she was "used to scraping and scrimping and making do patching things together . . . and you'd go in there, and it was all new materials and supplies. And the smells [of the wood] were lovely."

An older local businessman, Herb Badcock, who had lost his beloved wife Cynthia to cancer seven years earlier was another sponsor of the build. Not only did he purchase needed epoxy resin and bring expensive Godiva chocolates to the boat build sessions, but he also donated a gazebo that the Avalon Dragons used to raise funds through ticket sales and a drawing. Anne Marie described him as a "generous, kind man," adding: "He's sweet and doesn't want to see us wanting for anything, and I'd never had that in my life before."

The Avalon Dragons' position as the first group of breast cancer survivors in North America (and the second in the world) to have built their own boat caused Bruce to suggest to the women that they fashion a fiberglass mold from the completed wooden boat hull. The mold could then be used to produce future fiberglass dragon boats. The group readily agreed. Once again, the women focused on quality; Jane explained that the Avalon Dragons wanted to ensure that it was "perfect on the inside" so that it would easily "slip off the boat." She added that had the mold been built by the men at the university, which had been considered, "they probably would have just done it" without such care.

Racing and boat completion

By May, 2008 the Avalon Dragon Boating Association reached a critical decision juncture. The paddlers had planned several months earlier to participate in their first racing competition on August 7 at a dragon boat festival in New Glasgow, Nova Scotia, based on their belief that their wooden dragon boat would be completed and launched on July 5. But the wooden boat was being built carefully with the highest standards of craftswomanship, and the build was behind schedule.

So at the Association's regularly scheduled meeting on May 22, Julie presented the group with three voting options: 1) proceed as planned to get the mold built and finish the inside of the boat as quickly as possible, which would require an intensive building schedule (nearly daily), to which Bruce had committed; 2) abandon building a mold and work on finishing the inside of the boat as quickly as possible; or 3) build the mold, complete the inside of the boat under a normal schedule, and purchase a fiberglass boat for the New Glasgow race (it was mentioned that an organization called Great White North had a pink dragon boat available for purchase). After extensive discussion of the pros and cons of each option, the group first voted unanimously to remove option #2 from consideration, and then decided by majority vote to follow option #3, given that some donations had been pledged to help offset the purchase cost. Moreover, the official launch would be delayed until September, 2008. Not all of the women were pleased with the outcome for they had hoped to have their first dragon boat experience in the wooden boat that they had built. Still, it was decided that they would practice in the purchased boat without fanfare, saving all publicity for the launch of their hand-built wooden dragon boat in September.

The fiberglass mold

On May 26, Bruce and the boatbuilding women, along with husbands and other supporters, transported the completed dragon boat hull to Odyssey Yachts, a relatively new yacht manufacturing firm in Seal Cove (a 30-minute drive from St. John's) that had donated its space, equipment, and expertise for the making of the fiberglass mold. Several processes would be involved: positioning the boat, lightly sanding it, making a paper pattern for the corecell, measuring patterns for a needed cradle to support the boat when it returned to the boat shop, applying wax, cutting and hand-applying the fiberglass to the upside-down boat hull, and ultimately lifting the completed fiberglass mold off of the boat. The women decided that they were game for doing the fiberglass work themselves, despite their unfamiliarity with the work and the intensive schedule involved. Protective gloves and masks shielded them from the glass fibers and to some extent from noxious fumes, but the work was challenging nonetheless.

The culmination of their efforts would be the removal of the fiberglass mold, scheduled for Sunday afternoon, June 8, 2008. Neil Elliot retrospectively described the high stakes involved:

> That was a high-tension day. . . . if something went wrong at that stage and you had a year and a half put into your boat, and instead of the mold releasing from the wood it took part of the wood with it, it would have destroyed all you had worked for.

Two days before the planned removal of the fiberglass mold at a social gathering that Jane hosted to introduce me to the boat builders for later interviews, Bruce's ever-calm demeanor did not fully mask his concern about the potential of the boat to be damaged if the mold did not release from the hull perfectly and smoothly. Using a domestic work analogy, Pat Greene urged him not to worry, stating that the process was just like turning a freshly baked cake out of a pan, and that all would go well.

I was fortunate to have been in attendance that Sunday in Seal Cove, and the atmosphere was both festive and tense as the women boat builders, their husbands/partners, and Odyssey Yacht employees gently tapped shims underneath the boat to pry the edges of the fiberglass away from the wood. As they waited in anticipation of the mold lift, a number of the women huddled into a circle and composed a marvellous paddling song about their boat build, to the tune of the Newfoundland folk song "I's the B'y," beginning "We's the gals who built the boat; We's the gals who paddle her. . .". After a couple of hours, the moment of the lift arrived, and when the mold came off leaving the boat undamaged there were cheers and hugs of celebration among all present.

With the fiberglass mold completed, the boat was transported back to the Macdonald Drive Junior High school workshop in St. John's for the insertion of its seats and various other final touches. The dragon head and tail had been designed by Newfoundland artist Diana Dabinett, and the boat builders went to her studio in Shoe Cove to paint those parts of the dragon. An overwhelming desire to complete the boat caused Bruce and the women to undertake an intensive build schedule, thereby making great personal sacrifices. Pat gave up an opportunity to travel in July with her church choir to Beaumont Hamel in commemoration of the heavy loss of life the Newfoundland Regiment sustained there during the First World War. Many of the boat builders did travel to Nova Scotia in August to participate in the dragon boat festival, and their spirits were buoyed by their first competitive paddling experience.

The Avalon Dragon Boating Association had to work in the meantime to overcome a final hurdle – locating a lake for the September launch and a boathouse for year-round storage of the completed wooden dragon boat. Applications to the St. John's Regatta Committee for the use of their facilities at Quidi

Vidi Lake and then to the Pippy Park Commission and Memorial University for access to Long Pond were each rejected. Their last hope was to appeal to the nearby town of Paradise in which Octagon Pond was located. Julie Bettney stated that the mayor "was thrilled, absolutely delighted [at the prospect and] thought it would be a wonderful attraction for the town" which was in the process of developing a large, multi-use park around the pond. Permits were approved for a two-finger pier dock and a boathouse, both of which were built in a single month with labor donated by the Newfoundland and Labrador Road Builders/Heavy Civil Association. Julie estimated the total value of what was built to be about $130,000–$140,000, adding that:

> . . . anytime we do anything, we get all of this media attention . . . it's all positive – everybody feels good about it because . . . so many people are touched by breast cancer, and I think so many people feel helpless in the face of it, that to see the power, to see . . . what we've achieved, to feel like they're contributing to it in any way is beneficial.

The official launch of the hand-built dragon boat was set for Sunday, September 14, 2008, but the women, along with Bruce, their husbands, and other loyal supporters, transported it to Octagon Pond on July 23 in order to practice paddling and experience the joy of being in their boat for the first time. Neil Elliott was there to help and remarked: "When we moved the boat from the workshop" [to Octagon Pond] "the ratio of women to men that helped carry it out was probably 3:1." Marie described how protective the women felt toward the boat during the launch: ". . . none of us want to see her get scraped . . . we sanded [her] so many times and put so many coats [of stain and varnish] on. . . . We [didn't] want one little thing to be out of place; [we wanted] it perfect."

In fact, instead of breaking a bottle of champagne before putting the boat in the water, the women elected to toss birdseed and fish food in celebration of the launch, being sure to vacuum and clean the boat afterwards. Marie admitted that she and the other women realized that the boat would inevitably get scratched, and that in 20 years' time those who paddle her may not feel as concerned about scratches as the builders did.

The Avalon Dragons and their supporters were blessed with a mild, sunny day for their official launch of the dragon boat on September 14. All 50-plus members of the Association were in attendance, along with more than 300 cheering supporters who partook of the ceremony to "awaken the dragon" marked by drumming, tossing rose petals into the water, and painting red dots in the middle of the dragon's eyes (The Telegram 2008). The indomitable Pat Greene did not let the dislocation of her shoulder the night before deter her, despite late night treatment at the hospital emergency room; she attended the celebration, and drummed all day at the ceremony and in the boat while in pain, unbeknownst to the other Dragons. Reportedly her only concern was having her photograph in the newspaper seen by the treating physician (Brown 2008)!

Personal benefits of the build

Technical skills

The Avalon Dragons who were involved in the boat build experienced clear benefits pertaining to their use of technology. One obvious outcome was increased confidence in the use of tools and woodworking techniques. Janice, who described herself as a "pretty self-confident person" in general, said that she now felt more comfortable using unfamiliar tools. As a result, she was planning to help a friend build a kayak. June stated that she and her sister Ruth were considering volunteering to work on homebuilding projects with Habitat for Humanity. Pat stated that she is "more prone now to tackle" new projects, and in the latter months of the boat build she decided to remodel the master bedroom in their home and removed wallpaper, painted, plastered, and installed crown molding, adding "It was a real accomplishment and I don't think I would have tackled that on my own" without having partici-pated in the boat build.

Marie admitted to developing a fascination for machinery that she'd had no interest in before. However, she stated that she "loves working as a group" and wouldn't want to do "an individual build." Her husband Neil was con-templating building a dory; Marie stated that if he does so, she'd "love to go out and help him. I never would have thought that before." She added that:

> the whole process [of boat building] has made me a much more confident person, because I was a very shy and introverted person. . . . so all of these things [that] made me come out of the comfort zone have made me a better person.

It also changed her attitudes toward women in the skilled trades. Referring to the program with which Anne Marie had worked – "Orientation to Trades and Technology" – Marie used to wonder: "why would anyone want to do that? Why would anyone want to do something that was more male oriented?" After Marie's year and a half on the dragon boat build she observed: "a young girl who . . . was a most delicate-looking, tiny . . . little thing doing electrical work, and I thought – 'Wow! Good for you!'"

In a related vein, Jane described how she gained "a wealth of knowledge" about tools and their uses as a result of the boat build. Consequently, she was able to order a special tool sharpening device from a catalogue as a gift for her brother, which "worked beautifully for him and he was so impressed."

Mary described the benefits she experienced as follows:

> Physically being able to master the tasks of using these tools and knowing that I could create something, and basic carpentry skills, have hugely boosted my confidence in those areas. It's given me enough confidence to tackle jobs at home that I would never have thought of doing before – redoing a bathroom, doing some basic plumbing, building an extension to

my fence – I've done all those things in the last year just because I have the confidence and know where to go to get help and advice when I'm trying to do something like that.

Sylvia, who had not used power tools prior to the boat build, also commented on how the confidence she gained from that experience played out in a project in which she and her husband were involved:

> . . . as a result of using the tools over there, when we were clearing land down by the waterfront on our property, we had the chainsaw, and I said to Rog[er]: 'I can use any tool I want from now on, so get that chainsaw, show me how to use it.' And I loved it. I love hacking down trees with the chainsaw. So using all those different tools . . . [and] being taught the correct way and incorrect way to [safely] use them, and knowing that I had used the various tools over in the workshop to build the boat gave me confidence to say: 'Just show me how to use it; I can do it'.

Donna commented that even with her prior tool experience, her confidence had increased as a result of the boat build. Referring to tools that that her husband had in the house, she stated:

> . . . most of those now I can use pretty much with confidence, and the ones that I could use before, probably more so than ever simply because of the fact that I had all the practice. And the fact that I know that I was taught the right way, that I was taught by somebody who was very safety conscious.

Additionally, the women were more inclined to ask one another for help with homebuilding projects as a result of their boatbuilding experience. Such an appeal was made when the dragon boat was taken to Octagon Pond in July, and Marie's reaction was:

> I thought it fascinating because we can all go and help each other now. And that's what came out tonight . . . I was thinking that [in the past] we would never as women come up and say that. We'd say: 'Is your husband free to do it?'. . . . [now] we would not even think twice about it. We would just say: 'Where's the job? Let's go!'

Health benefits

Another positive outcome was the opportunity to improve one's physical fitness – thereby boosting self-confidence and health. Sylvia stated:

> I certainly would not be a gym member if it had not been for the Avalon Dragon Boating. . . . that is one of the great things that has come out of

this boat build. . . . I have been active most of my married life in that I walked every day; I worked out with weights in the basement. But I've never participated in the gym and in classes, and the boat build has done that for me.

June also appreciated the physical aspect of her association with the Avalon Dragons, as well as the making of new friendships. She had cared for her mother in her home, and spent a great deal of time there. Learning how to paddle and going to the gym were new experiences for her, and as a result of her involvement with the Dragons, she hoped to "get and stay in shape." June also remarked that the praise she received from Bruce and "the other girls" bolstered her confidence. Her husband Don commented on the "big, big change" in June over the period of the boat build, especially "confidence in herself" and a recognition that one can "do anything in life if you put yourself into it."

Interpersonal support

The boat build sessions offered participants a supportive environment in which to immerse themselves while dealing with difficult health issues and other stressors. The women also benefitted from getting to know one another and the camaraderie that developed among them. Pat enjoyed "meeting new people, and making new friends." Ann did as well, and beyond friendship she enjoyed:

. . . being around people. You weren't talking about cancer; you were just talking about your own experiences. . . . It was stress relief because . . . [close family members] had some medical problems. . . . It was enjoyable just to go and your mind was relieved.

She characterized the Avalon Dragon Boating Association women as "inspirational" and "very supportive of each other." As the longest breast cancer survivor among the boat builders (21 years), Ann revealed that the group:

. . . motivated me to be a stronger person, to want to do more with helping people, to getting the message out there that there can be life after cancer, and I think that at the end of the day and once we get our festivals going that people are going to have a different outlook on cancer in general. It's a joy to be part of our group. It is an inspiration. It's something that you want to do. . . . Everything is stopped just for the dragon boat [build] . . . it's something that affects my heart, and something that is fulfilling at the end of the day.

Janice appreciated developing a network of people to call upon for information, and "experiencing the dynamic group of Avalon Dragons" whom she described as a wonderful cross section of different individuals of different ages, and very focused and respectful of one another. Mary observed that the group served as a great source of support for those members who had lost jobs, or who had faced the deaths of parents and the illness of spouses. The women also offered empathy and understanding to those who were in breast cancer treatment (she herself was a 20-year cancer survivor). However, like Ann she indicated that the group didn't talk about cancer per se, but rather talked "about everything else."

Jane, who was involved in three committees – sponsorship, boat, and paddling – and who attended all of the boatbuilding sessions, stated that the build was "stress free, conflict free and a place of refuge for so many after death in the family or losing a job." Anne Marie appreciated "being surrounded by all this positive energy . . . from women that I really admire." She added: "I really think what I've gotten from this organization more than anything else . . . [is] strength – I've become a strong woman. I've come into an environment where I'm allowed to be strong."

Sylvia described a number of benefits she derived from her boatbuilding experience, friendship and learning foremost among them:

> One of the nicest things that's happened to me at this stage of my life is to have gotten involved in the boat building. I have made some new friends. . . . You go over to the boat build, and you're there at any given time with 8, 10, 12, 14 ladies, and it's all fun and laughter. You don't hear everyone moaning and groaning about what kind of a day they had yesterday, or about their illness. Cancer is not a part of our lives. Now every now and then someone might strike up a conversation depending on what's on the go, may talk about some aspect of . . . [her] life with cancer. But that's not what's over there in that boatbuilding.
>
> And the same with paddling. . . . You don't have any depressing people around you. It's unusual that you have so many women together and they're such beautiful women. . . . they are just nice women, and they're so gifted. . . . it's amazing the wonderful gifts that so many of those women have. The gift of fundraising; the gift of making things happen. It's just incredible. The women are all so pleasant, and they come from all walks of life – they're just really beautiful women to be around. It's one of the nicest things that's ever happened to me at this stage of my life. I have been so blessed that I have become a part of this organization, of this group of ladies. . . . and it just keeps on unfolding – now we're in the boat; now we're paddling; now we're learning something new – all the new things that we're learning out on the water. . . . and we'll go on to a festival; we will meet many more beautiful women, and we will learn many more new things. How lucky am I? How blessed am I? I always say that cancer has opened many new doors for me. It's a wonderful teacher, but a hard one too.

Other benefits

It is clear that although the women were learners in the boatbuilding experience, they were also in a certain sense teachers. Their courage in overcoming the physical and psychological scars of breast cancer in such a public way served as an inspiration to many in the community and beyond. Spouses/partners, children, siblings, and other family members were all too happy to support the women's boatbuilding, paddling, and other Dragon Boat Association activities – all of which required many hours away from home, for they saw the transformational effects those experiences had on the women they loved. The boat builders also served as mentors by demonstrating to girls – like the 30 Girl Guides who visited the boat build on February 28, 2008 – that women could indeed master tool use and construction techniques and work together cooperatively.

It was obvious that Bruce found his work with the women boat builders deeply moving. His extraordinary commitment of time and expertise overlaid with patience must certainly have been encouraged by their lessons of fortitude, resilience, and joy in the face of adversity, and the community of friendship that evolved among virtual strangers. That community extended beyond the confines of the MacDonald Drive Junior High industrial arts room, as mentioned earlier, but there was a watertight energy bond at the boat build that was as durable as the epoxy used.

The experience gave Bruce, who came from a family of physicians, the opportunity to promote healing in a unique manner, using a formula that blended boatbuilding and design passion with humanity. As his brother John – himself a medical doctor – observed after a visit to the boat build, Bruce "patiently acts like the collective's 'physician' – listening, analyzing, identifying problems, offering solutions, complementing, and treating. There is no doubt that woodworking can be therapeutic, but what's going on in this old woodworking shop is magical" (Whitelaw 2008, 27). Bruce was not about to let the completion of the boat terminate his intensive involvement with the Avalon Dragons; he thereafter took on the role of paddling coach.

Among the most profound lessons of the boat build were those conveyed by the words and actions of Donna Howell. Although she never mentioned it, I learned that she was one of the women who had been misdiagnosed in a laboratory scandal that shook Newfoundland (Sullivan and Woodford 2007; Adhopia 2008). As a result, the breast cancer that went undetected for some time ultimately invaded her bones and liver. Ever upbeat, Donna described what the boat build had meant to her, and expressed determination to paddle at a competition in the coming year:

> I knew that I would enjoy [the boatbuilding] because I like that kind of thing, but I didn't know how much, and I didn't know how many friendships I would form as a result of all this. And how important it would become to me, and it has become very, very important. . . . It's certainly an experience that I would not have traded for anything. . . . I know what the prognosis of my life is, but knowing that I've got all these people there who are doing well, and we're all fighting [is heartening]. I couldn't be in the boat to go away this year [for the August race in Nova Scotia], but I plan to be there next year!

Donna's association with the other boat builders gave her confidence and strength, and she in turn passed that along to others by virtue of her life-affirming spirit:

> I think that just seeing all of those women who have had breast cancer . . . alive and working well, and strong and healthy and positive in their attitudes gives you a lot more confidence in what you can do yourself. I've always been fairly optimistic; that's been part of my nature anyway, but because I have two girls . . . you have to set a positive example because the possibility is always there that one of them could develop this at some point in time. And I don't want them ever to have the idea that once you've been diagnosed life is all over, because it can't be like that. You have to get up and you have to stand up and fight and make every day count. And even though I always felt like that, seeing all those women and everybody doing well and thinking positively [was helpful] . . . and we all have our times too when you get concerns and something comes up, but you knew that you could just go there and you might have told one person ahead of time, like when I got diagnosed with the liver cancer. So when I told Jane then, she just told the rest of the girls [who] were there . . . they didn't say anything; they just knew, and you didn't have to . . . rehash it, and yet you knew that they were all there for you.

Donna's appearance at Seal Cove on June 8 to witness the removal of the fiberglass from the boat was the first time she had participated in boatbuilding activity since January 2008. Her treatments had kept her away from the boat build for those five months because the chemicals used for the varnishing and other processes aggravated her reaction to the chemotherapy. Yet, she found it therapeutic to visit when she could:

> . . . in between that I needed to just go and see them every so often . . . I just feel better when I come away. You just knew that they knew everything probably that you were going through, and that they understood, and they had done it, and you just felt that you could do it . . . that they believed in you, and so therefore you believed in you.

Sadly, Donna passed away on September 25, 2009 at the age of 53 (CBC News 2009).

Postscript

The wooden dragon boat built by members of the Avalon Dragons is not raced, but rather paddled each year for a very moving "Flower Ceremony" at the annual "Paddle in Paradise" Dragon Boat Festival at Octagon Pond. "Paddle in Paradise" is a celebratory event that serves as a fundraiser for the Avalon Dragons in which breast cancer survivors and many other community members of both genders are sponsored by local businesses to race in fiberglass dragon boats. The Flower Ceremony commemorates those who have passed from breast cancer, and the Avalon Dragons paddling the hand-built wooden boat toss pink rose petals into the pond, guided by the drummer and steersperson. I attended that ceremony on July 28, 2012, during which a recorded song was played over the loudspeaker: "Rolling River" by Shari Ulrich, with a fitting chorus:

> There's a rolling river to carry us home
> We are cold and weary
> But never alone
> Made of tears and courage
> Flesh and bone
> Oh we built this boat to carry us home
> (Ulrich 2010)

Some of the original boat builders have been involved in annual competitive races outside of the province, including the International Breast Cancer Paddlers' Commission Participatory Dragon Boat Festival held every four years in a different country or province. In 2014 it came to the U.S. for the first time, and was held in Sarasota, Florida from October 24–26 (IBCPC 2014). Two Avalon Dragon teams competed well; one "crossed the finish line in first place" and another "finished the day in the top third of over 100 teams racing from eight countries, including Canada, the U.S., England, Ireland, Italy, Australia, New Zealand, and Singapore" (The Telegram 2014, 1).

Conclusion

The Avalon Dragon boat builders were motivated by their desire for improved health, connection with other breast cancer survivors, and ultimately by a desire to engage in a collective project resulting in a product of beauty and utility. Mentored and taught by a wonderful male instructor, they also engaged in peer learning, and took pride in the quality of their work. They were encouraged by supportive family and community members. The confidence they derived from their successful woodworking and mastery of tools carried over into other aspects of their lives. They were pioneers in breaking the cultural tradition of boatbuilding as a masculine endeavor.

The finished wooden boat is clearly a work of art. Yet as the stories conveyed here demonstrate, it also represents a work of heart and stands as a testament to the power of positive collective energy to heal and transform.

Note

1 Newfoundland and Labrador together are a single Canadian province. Since the boat build took place on the island of Newfoundland, and since the Avalon Dragons were based in the provincial capital of St. John's, I will at times refer only to Newfoundland in this manuscript.

2 Digital megaphone

Egyptian women's cyberactivism in the Revolution and beyond

Women have been centrally involved in every one of the 21st-century political uprisings seeking democratic governance in Arab countries – most notably in Tunisia, Egypt, Yemen, Libya, Bahrain, Syria, and Saudi Arabia (Radsch 2012; de Silva-de Alwis 2011; Morgan 2011). Yet while the "Arab Spring has been a lightning rod for some important reforms on behalf of women" there is a compelling need to "preserve prior gains . . . and to ensure that women are at the forefront of transitional justice in this period of historic transformation in the region" (de Silva-de Alwis 2011, 4). That is especially so in Egypt, where women were politically marginalized in the constitutional reform process that followed the overthrow of President Hosni Mubarak and treated harshly by the Supreme Council of the Armed Forces (SCAF).

Egyptian women played a significant role preceding, during, and following the Egyptian Revolution of 2011. They fought for an end to government repression and corruption, and also for democratic reform and women's rights. This chapter highlights the ways in which six Egyptian women activists who were involved in that Revolution mastered and used information and communication technologies (ICTs), particularly social media tools, in support of their quest for social change. It is based on interviews conducted with them in the fall of 2012 and subsequent e-mail correspondence. Some contextual background information on ICT use in Egypt prior to the Revolution and on Egyptian feminism is provided before presenting that case-study data.

ICT growth in Egypt prior to the Revolution

An Egyptian professor wrote in 2007 that while the "Arab world" has historically been "on the low end of the digital divide . . . the Internet growth rate in the Arab world is exploding" (Abdulla 2007, 35). Yet Egypt was at the center of information dissemination thousands of years before the Internet made its debut:

> Egypt, through its ancient history which extends over 3000 years BC has been witnessing massive information flows through different means since the era of the Pharaohs. . . . life and development along the Nile was paralleled by a different type of an information society. This included

> inscription on stones, papyrus papers, Rosseta stones and the establishment
> of the library of Alexandria which was considered in ancient Egyptian his-
> tory, the first largest and famous library world-wide. During the middle
> ages, Arabic manuscripts, documentation on Papyrus together with doc-
> umentation on "Parchemin" which is a form of leather became one of the
> most common means for information dissemination. In the modern age . . .
> printing and publishing of the first Egyptian journal [occurred] in 1826. A
> few years later in 1830, Egypt witnessed the establishment of the first
> national archive system.
>
> (Kamel 1997, 8)

Development of the Internet in Egypt dates back to 1993, initiated by the
Egyptian Universities Network and France (Kamel 1997). By 1996, Egypt had
the highest rate of Internet connectivity of any Middle East and North Africa
(MENA) Arab state, thanks in large part to public policy supporting Internet
growth, including permitting private-sector Internet service providers (ISPs) to
own and manage their networks (El-Nawawy 2000). The government recognized that an information highway was essential to socio-economic growth, and
in 2002 the Ministry of Communications and Information Technology in partnership with private ISPs launched the Free Internet initiative, thereby providing
subscription-free dial-up Internet for all Egyptians (Egypt Ministry of Communications and Information Technology 2009). Mobile phone communications
had also grown after 1996 when Egypt Telecom introduced Global System for
Mobile Communication (GSM) services (Kamel 2004). A decade later, Egypt
was the third largest mobile market in Africa after South Africa and Morocco
(Hassanin 2007).

As in all countries, ICT availability has been uneven, with "digital divides"
related to geographic location, income, education, gender, and age group (Arab
Republic of Egypt 2015). Literacy has been one barrier to Internet use among
certain segments of the population. Prior to the Revolution, female adult literacy (over 15 years of age) was 63.5 per cent in 2010 compared to 80.3 per cent
of males in the same age group; the rates were higher for youth 15–24 years old:
90.6 per cent for males and 84.3 per cent for females (UNESCO 2011). In an
effort to improve those rates, UNESCO's Regional Offices and Cairo and
Beirut, in cooperation with the Egyptian Ministry of Education, launched in
2011 a National Campaign for Literacy and the Renaissance of Egypt initiative
entitled: *Together We Can* (UNESCO 2012).

Internet censorship and repression

Access to the Internet and social media devices that broadcast through it have
not necessarily translated into online freedom of expression. The OpenNet
Initiative reported in 2009 various ways in which the Egyptian government had
stifled anti-government dissent and increased Internet surveillance, arresting
people who had used cell phones for calls and text messaging to help organize

demonstrations (OpenNet Initiative 2009). The same report mentioned that the Committee to Protect Journalists named Egypt one of the ten worst countries in which to be a blogger.

ICT censorship intensified during the height of the Egyptian Revolution. On the eve of a major demonstration planned for January 28, 2011, the Egyptian government prevailed upon ISPs and cell phone providers to shut down their networks, resulting in a 90 per cent drop in data traffic shutdown as "unprecedented in the history of the internet" – not only in its scope but also in its crippling of an anti-censorship tool that activists had used to circumvent earlier blocks to Twitter and Facebook (Williams 2011, 1). The demonstrations went on nonetheless, thanks to word-of-mouth and older technologies like fax machines, ham radios, and dial-up modems using international numbers, until Internet access was restored days later (BBC 2011a; 2011b). Such resourcefulness was laudable since social media sites like Facebook and Twitter had been used to initiate and promote the Egyptian Revolution during its early phase (Zeid and Al-Khalaf 2012).

The resignation of President Hosni Mubarak unfortunately did not end the harassment of pro-democracy activists using the Internet. The SCAF continued its campaign of repression by interrogating, arresting, and imprisoning bloggers; women activists were not exempt from such maltreatment, as Bothaina Kamel, Mona Eltahawy, and Asmaa Mahfouz, among others, discovered (Reporters Without Borders 2012). Women played a central role in the protests that toppled the government, fighting not only for democracy and human rights, but also for gender equality.

Feminist activism in Egypt historically

Women who poured into the streets of Cairo and elsewhere during the Egyptian Revolution followed a long tradition of feminist activism in that country. Huda Sha'rawi, Nabawiyya Musa, and Malak Hifni Nasif (also known by her pseudonym Bahithat al-Badiyya) were among the earliest figures who publicly engaged in the struggle against patriarchal norms limiting women's independence and education, and simultaneously against colonial rule (Badran 1992; Badran 1995; Yousef 2011). The education of women was seen by liberal nationalists of both genders as essential to the advancement of a post-colonial Egyptian nation (Badran 1995). Sha'rawi, an upper-class woman, organized public lectures given by Musa, Nasif, and other newly educated middle-class women, and helped to establish the Women's Intellectual Association (Badran 1992, 35). Sha'rawi and Musa became founding members of the Egyptian Feminist Union in 1923, five years after the untimely death of Nasif (Badran 1988). Those early Egyptian feminists recognized that feminism and Islam were compatible:

> Women's feminist analysis began with the process of disentangling patriarchy and Islam. Women discovered that the veil was not required by

Islam, nor were sex segregation and female seclusion. They also realized that Islam guaranteed women's rights that patriarchy withheld from them. At the same time, they claimed that women's advance would benefit the nation displaying a feminism with a distinct nationalist dimension.

(Badran 1988, 13)

Writing from a feminist perspective born of her Egyptian childhood experience and the lessons learned from elder women in her life, Ahmed (1999, 2012) explained what gave rise to different interpretations of Islam:

> . . . there are two quite different Islams, an Islam that is in some sense a women's Islam and an official, textual Islam, a 'men's' Islam. . . . Mosque going was not part of the tradition for women at any class level (that is, attending mosque for congregational prayers, . . . as distinct from visiting mosques privately and informally to offer personal prayers, which women have always done). Women therefore did not hear the sermons that men heard. . . . They did not have a man trained in the orthodox (male) literary heritage of Islam telling them . . . what it meant to be a Muslim. . . .
>
> Rather they figured these things out among themselves. . . . as they tried to understand their own lives and how to behave and how to live. . . . and as they listened to the Quran and talked among themselves about what they heard. . . . For this was a culture, at all levels of society and throughout most of the history of Islamic civilization, not of reading but of the common recitation of the Quran. It was recited by professional reciters, women as well as men, and listened to on all kinds of occasions — at funerals and births and celebratory events, in illness and in ordinary life. . . . The dictum for that 'there is no priesthood in Islam' – meaning that there is . . . no need for an intermediary or interpreter between God and each individual Muslim and how that Muslim understands his or her religion – was something these women and many other Muslims took seriously and held on to as a declaration of their right to their own understanding of Islam.

(123–125)

Even though Sha'rawi's public casting off of her veil is legendary, Badran explained that:

> For complex reasons there was a lag in Egypt between Muslim women's discovery that the face veil was not required by Islam and their discarding of it. Even if, strictly speaking, covering the face was not ordained by Islam, the association of the practice with the religion was compelling, and was bound up with issues of identity, security, and respectability.

(Badran 1992, 37)

Thus, Sha'rawi "favoured gradualism" since:

> . . . the feminist vision was concerned with a longer process of liberation
> which they did not wish to threaten. The feminists did not want women to
> become unnecessary victims through assaults on themselves and their
> honor. . . . They insisted upon their own agenda and timetable.
>
> (Badran 1992, 39–40)

British colonial rule was also a source of patriarchy that feminists like Nasif
recognized as distinct from nationalist patriarchy (Yousef 2011, 82). Once Egypt
attained nominal independence from Britain in 1922, "the moment was right for
the feminists to move to desegregate society through visible, public activism to
further their ultimate feminist goal of full liberation for women" (Badran 1988,
18). Yet their active participation in the nationalist movement did not confer
upon them true equality. As Hatem (1992) noted, personal status laws that were
passed in the 1920s and 1930s ". . . defined women as the economic dependents
of men, unstable emotional beings that cannot be trusted with the right to
divorce, and unable to leave a husband without his consent and/or in cases
where he is incurably ill or impotent." And although women were granted the
right to vote in 1956:

> . . . state feminism under the Nasser regime produced women who were
> economically independent of their families, but dependent on the state for
> employment, important social services like education, health and day care,
> and political representation. While state feminism created and organized a
> system of public patriarchy, it did not challenge the personal and familial
> views of women's dependency on men that were institutionalized by the
> personal status laws and the political system.
>
> (232–333)

In 1979 President Anwar Sadat passed decrees changing the personal status laws
and assuring women 30 seats in the Assembly and 20 per cent of all seats in
local People's Councils. Hatem (1992) contended that those laws were not
indicative of state feminism's survival, given "the lack of a coherent state program on gendered issues during this period." She added:

> Unfortunately, the state's desire to continue to dictate social and political
> policy made a mockery of official rhetoric regarding political liberalization
> and further undermined its legitimacy. It made the Sadat regime doubly
> inadequate. Not only did it fail to provide economically for its own citi-
> zens, but in addition it continued to be politically authoritarian.
>
> (243)

The Revolution of 2011 occurred against this historical backdrop of authoritarian rule and internal divisions, as described by Talhami (1996): "The

struggle for women's rights in Egypt, as elsewhere in the Arab world, is a continuing battle affected by sectoral divisions within female ranks, indifferent seclar men, and authoritarian rulers schooled in cooptational politics" (146). Yet modern activists could take inspiration from the legacy of courageous women like Mounira Thabet, Nabawiyya Musa, Siza Nabrawy, Doria Shafik, Nawal el Saadawi, and others who fought for women's rights – often at great personal costs (Osman 2012; Al-Ali, 2002).

The contributions of six feminist cyberactivists to the Revolution and beyond

This chapter offers original case-study research on how six Egyptian women – all of whom identify with the term "feminist" – used the Internet to advance their political objectives on behalf of women's rights, human rights, and democracy during the Revolution of 2011, and also before and after that event. In keeping with the purpose of this volume, the questions asked of interviewees focused on the factors that enabled them to master and use the Internet and social media tools in ways that were personally and collectively empowering. The interviews were conducted in the September and October of 2012 by telephone and by Skype in the English language – some 18 months after the uprising. Although much has been written since the year of my interviews about Egyptian women's role in the Revolution (Allam 2017; Mostafa 2016; Baker 2015; Herrera 2014), this chapter is an in-depth snapshot in time of six feminists and the tools they used to promote social and political change.

Demographic profiles

The six women ranged in age from 23 to 70+ years at the time of the interviews. Their activism took varied forms, with a common theme of working for justice, equality, and democracy. The following descriptions of their activities in 2012 continue to be accurate unless stated otherwise. "Hasiba" – a former university professor at Cairo University – heads an Egyptian feminist NGO and serves as chair of another Egyptian women's organization for individuals; she stopped teaching to devote all of her time to those organizations. "Bayan" works in mass media and contributes her energies to the promotion of democracy and gender equity in national electoral politics. "Sara" co-founded an organization that does sexual harassment prevention throughout Egypt; her paid work was in international development, assisting women's rights NGOs focused on family law and violence against women. She presently works for an international development agency.

"Hafa" also does educational volunteer work on sexual harassment prevention in Egypt and works as a journalist, formerly for an international progressive news organization and presently as a foreign correspondent for mainstream news organizations in two other countries. She has a Bachelor of Arts degree in International Law and Legal Studies from Cairo University.

"Dalal" was and remains a women's rights activist doing volunteer and paid work on gender equality and women in politics in Egypt and regionally in the MENA. At the time of the interview she was in the early stages of writing a doctoral dissertation in Political Science and Iranian Studies at Cairo University on women's political participation in Egypt, Iran, and Pakistan during periods of structural transition.

"Durar" had worked for the "Arab Women Organization" (AWO) branch in Cairo. The AWO is headed by the "First Ladies" of all countries affiliated with the League of Arab States (Arab Women Organization 2016). In describing her work with the AWO, she recalled:

> . . . the one thing I really appreciated and I really loved was . . . being exposed to women from 16 Arab countries – meeting women from Iraq, from Palestine, from Amman, from Morocco, from Tunisia, I met women from Mauritania for the first time in my life, so that was very interesting for me also.

Durar had also written for news and political affairs magazines published in English and Arabic before completing her Master's degree in Gender Studies at the University of London's School of Oriental and African Studies (SOAS). While in London, she had worked as a consultant for a London-based organization working to advance the rights of women living under Sharia Law in Iran, Pakistan, and Africa. As of 2018 she was completing her doctoral studies in Germany on the Egyptian Revolution and has published on the topic of gender in that Revolution.

In summary, all of the women were well educated, and had earned their undergraduate degrees in Cairo (most at Cairo University) in the following fields of study: political science/social sciences, journalism/mass communication/media studies, or international law. Three of the interviewees had completed or were pursuing graduate degrees in Egypt or abroad. Their parents were also university-educated, with the exception of Hasiba's mother, and some had graduate degrees.

Use of social media tools in political work

Benefits and user preferences

Each of the interviewees, irrespective of age, reported using various forms of social media in their political work. In response to the questions: "Have you used social media technologies to advance your political goals?" and "Which social media tools have you found most useful?" Hasiba, the eldest of the interviewees, replied in connection with the women's organizations with which she was working: "Of course. A lot of communication is being done through the 'Net and through all kinds of social media. We have a Facebook [page], we have Twitter, we have Websites . . . and we have e-mail [addresses]. . .. We use all of them."

Dalal explained the value of social media technologies to activists, while recognizing the potential for abuse of the very same technologies:

> Some people use them for social issues, but for activists, these tools are brilliant because . . . it helps you to learn what is happening around you in the world . . . and then to connect with people to see how they can help you with your cause, and on a very personal level . . . the feeling that some person in the world really cares about what is happening to someone they have never met, but they are willing to write petitions, sign them, go to the administration, just to support people they've never met because of the social media. It's helping people to connect all around the world. Other people are using it to spread violence and hate speech. . . . It's a good and bad tool, as anything in the world.

While the interviewees characterized all of the social media tools as being useful in their political work, they recognized that certain tools are better suited to a particular purpose than others, and some of the women indicated their personal preferences. Bayan agreed with Hasiba that:

> All of them . . . every one of them is important. We can't say that some media [are] more important than other[s] because every one of them has their own audience or viewers or people who follow [them]. . . . I use all of them. . . . We started by Facebook at first to make [publicize] events, but through the Revolution and after the 18 days in Tahrir Sq, Twitter became important, because it's like a news agency . . . you can say what you want to say to people.

Dalal described the different purposes for which she uses particular social media tools:

> Twitter is most useful when doing political advocacy and campaigning – it's the fastest and most reliable tool. I use Facebook for a closed circle of friends and family; but Twitter is better for exchanging information and mentioning people and sending opinions . . . and it's very fast, it's on my mobile, so in 140 characters I can send that communication to various people around the world. I use Facebook for social campaigning and spreading information on women's issues among family and friends.

She reported using Twitter:

> . . . to send news out; I tweet about conferences I attend, . . . events, demonstrations . . . and to spread news and comment on . . . articles and opinions written in the media . . . I use it for women's rights and exchange information with people from different places in the world. Sometimes I meet people on Twitter and then years later I meet them in person.

Hafa, too, described her nuanced use of social media tools:

> Twitter – both in my work as a journalist and activism is possibly the most
> valuable tool because it's the best way to find out about things as they
> happen; it's an excellent way to share information, it's an excellent way to
> broadcast about your cause and to market your cause; it is a very quick
> way. . . . Facebook is better for organizing events; you can start an event
> on Facebook, you can start a Facebook group for talking about issues as
> well. Twitter is much faster.

In response to my question about whether she finds the word "limitation"
constraining, she added: "I used to find it very constraining; it's a matter of
getting accustomed to the medium more than anything else, and now that you
can link (to articles, etc.) it's much easier to do things."

Two obvious benefits of Twitter and other ICT tools have been: 1) activists'
ability to quickly transmit information – including re-Tweeting messages sent
by others who lack a large online following, and 2) the users' ability to bypass
government constraints on information access – a point conveyed by Bayan:

> . . . in April of 2011 . . . I was with the Revolution and we had a lot
> of problems – the official media and the TV is under control [of the
> government] and we couldn't talk openly about things. We can't say
> that social media and Internet is enough; but if you have it, it's an
> option . . . [beyond government-controlled traditional media]. . . . The
> social media – Twitter, Facebook, or Blogs or You-Tube is an option.
> It's become more popular day by day. I remember when I used Twitter
> before the Revolution, Twitter was for the elite, but when the [social]
> media became more popular and ordinary people entered, not only
> activists, [it became more useful].

She added: "Now our newspapers use news from Facebook and Twitter by the
citizen journalists" [on their mobile phones]. "It becomes like a fashion, and
people become curious" [and want] "to know about Twitter and Facebook."
She referred to the value of social media following the brutal killing of Khaled
Said in Alexandria, stating that "Facebook was the most important media"
source in spreading information about his murder, since initially there were no
television news reports about the incident.

The interviewees' familiarity with the various forms of ICT available gave
them deliberate control over the use of those tools. Durar offered examples of
her preferences for various applications:

> I used to blog, but I don't anymore. I don't Tweet; I have a Twitter
> account but I don't use it that much. It was mainly blogging and Facebook.
> I used Facebook heavily, but I don't Tweet that much; I have friends who
> do. Sometimes I use Twitter to know the news, because it's much, much

faster in Egypt than regular news and sometimes even faster than Face-book. . . .

You decide how you use it. So your Facebook could be some place where you just have a laugh with your friends, or it could be something serious, or it could be both. And Twitter is the same. It could be a way to meet people and I know a lot of friends get to know other friends through Twitter, or it could be a place where you find news, or it could be whatever you want. I can't tell you that Twitter is for this and blogging is for that. You decide really what you make of it.

She also left the door open for future blogging: "I think really when I have stories to write . . . whenever I have something to write, I blog . . . if I have more stories to write I might blog again. I haven't decided to stop forever."

Limitations of social media as experienced by the six cyberfeminists

Technology access

While the activists clearly recognized the benefits of social media as a digital megaphone, enabling them to project their messages to local and global audiences, they were also well aware of the limitations of such. Hafa indicated that access is hampered without the right technology:

I didn't have a smart phone until recently, so I could only access Facebook and Twitter on my computer, which was slightly inconvenient, especially if you want to know about things quickly . . . as an activist . . . you want to be on the ground and doing things.

Hasiba experienced technology limitations in relation to hardware and connectivity: "There are a lot of obstacles and problems dealing with technology. Sometimes the 'Net closes – we cannot continue working; sometimes the computer just . . . hangs up . . . It's a constant irritation."

Questionable accuracy

Hafa emphasized that as with any media form, the accuracy of what's posted on social media sites must always be checked:

. . . the problem with Twitter is you can't be entirely sure about the veracity of people reporting . . . you have to take it on trust to a certain extent, which you do I guess with any kind of press or any kind of media . . . You have to take everything with a pinch of salt and figure out if it's true or not.

[During the Revolution, an event posted] sounded so much more dramatic than it was or is; and then you go down and it is pretty bad but . . . [may not be] as dramatic as it sounds.

Another problem, as Bayan explained, was infiltration of social media by the police:

> The most important thing is to fight the police and the controls, because they use also Twitter. I discovered that some people around me use Twitter and they are from the police. And they want to break the credibility of the activists.

Hacking and surveillance

Security and privacy risks were mentioned by Durar as an obstacle to social media use. She stated that she had been the victim of hacking of her Facebook profile, and a false profile was created under her name. While the problem of identity theft is a common one globally, she also raised the specter of surveillance of social media sites:

> I don't want to be too paranoid, but I remember that the time of this happening that was problematic . . . was a few months after the Revolution with [the emergence] of the military council . . . a friend of mine told me it could be the government, but then again the government probably knows everything there is to know about us anyway. . . . Also, a lot of people get hacked every day from corporations that use their profiles to spam others. . . . It's not only government.

Bayan cited evidence of government and military monitoring of social media sites, noting that transcripts of Tweets made by activists were being introduced as evidence against them in military trials, "And now [in 2012] we have people in prison because of their comments on Facebook insulting the President." She explained that the repression of free speech that occurred during the presidency of Hosni Mubarak, and then by the ruling SCAF, continued following the election of President Mohamed Morsi. "We are in the same regime, the same system. They killed the Revolution, they stole the Revolution. . . . I have the hope always that we have gone the first step and we must continue." Beyond outright prosecution of activists based on their social media postings was their harassment by those believed to be paid commentators. Bayan stated:

> . . . the most ugly [aspect of] Twitter and Facebook are the electronic commentators who insult you all the time very badly. We have people paid by organizations, by the SCAF, . . . by I don't know who, by security, and they work only to insult you. And it's not good.

Dalal, too, bemoaned the negative messages posted to her Twitter account:

> What has been really problematic . . . is people entering into infinite arguments about the issues I talk about like women's rights and feminism –

issues that people will find very controversial in that context, like when I talk about violence against women, marital rape, FGM – female genital mutilation, or abortion. . . . They start bombarding my account and e-mail, and [say that] . . . this is an Islamic society, and these issues shouldn't be raised. . . . I try my best not to go into this type of endless discussion . . . I focus on a human rights perspective, and I'm sticking to it. And I'm not going to argue about religion or beliefs . . . because I believe religion is a very private issue. . . .

And I don't think the online tools – social media like Facebook and Twitter are really created for deep, sophisticated discussions, but rather for sending short messages . . . for [conveying] sending specific information – news and to connect with people. . . . If we want to go into deep discussion there are offline places like conferences, and workshops and seminars. . . . it's best to ignore [negative comments] and not take them personally, especially if they come from people you don't know.

Harassment and contention

It is not only vitriol from unknown sources, but also negative comments from people known to the interviewees that caused them to characterize the Internet as a contentious place at times. Durar revealed that she is quite open with her Facebook profile; her pictures are there for all of her friends to see. And she stated that could be a problem because she has 1,000 people on her Facebook page:

Some of them are family members or distant family members . . . we clash most of the time and we have different ideas for everything. Most of the time I get judged by the pictures I post or the opinions I write, and some-times I get negative comments. Sometimes I have to think very hard on who I accept, but I cannot *not* accept these people because they're either family or acquaintances. . . . I try to be very open about who I am.

During the Revolution . . . I had to delete some of my very close friends . . . from my friends list because we started fighting about the Revolution and it was very emotional because we were going through a lot and I couldn't handle them saying bad things about people who were dying . . . I really couldn't handle it; I couldn't see their posts; I felt that I really didn't want to know these people again. But after that we became friends again. It was such an emotional time, and they were being offensive [about the people who died]. It became a matter of not different opinions; it became a matter of life and death.

She was able to resume her friendships with them after the intensity of the Revolution subsided: "A lot of people did change their minds but others didn't; I just managed to . . . [recognize] that it's not a matter of life and death any-more. We can say different things."

Simplistic interpretations by Western media

A concern expressed by Dalal was the way in which Western media romanticized and simplified the Revolution, attributing its success to the use of social media by young activists. Although she indicated appreciation for Rich Site Summary (RSS) news updates from sources like CNN, Reuters, and AP, she also recognized the limitations of international reports on what she characterized as the "so-called Arab Spring" because they described a

> . . . utopia of what is going on in the region, which is romanticizing the issues . . . [with] incorrect assumptions, and they came up with the famous question: 'Where are the women in the Arab Spring?' The women have been there from before the Spring and during the Revolution and continue. But the international media is looking for all those good looking young women, and they ignore the trundle of women which has been there for ages. . . . [not only the famous women like Nawaal el Sadawi, but] all these unknown women who have been struggling for freedom and justice and equality for many years . . . ; they can be factory trade union leaders, they can be politicians, but the media chooses to ignore them because they don't fit into the utopian view. The women's movement is documenting the stories of these women . . . so that other women can identify with them and will be more willing to join with them in the struggle. The women's movement is writing what we call 'herstory', not just history.

This was a message that Durar was also keen to convey:

> I'm always worried about putting too much weight on social media. I'm not denying that it was a tool and still is a tool. . . . But sometimes in some articles . . . or at some of the conferences, they simplify the whole struggle, calling it the 'Facebook Revolution' and they really narrow it down. . . . But I know that 40% of Egyptians are illiterate and I don't know how many of them are computer illiterate; so I don't really know what is the Facebook or computer penetration at all in Egypt, but I'm guessing they're not huge numbers. This fascination with social media . . . is almost trying to make the Revolution look good or tech savvy or fashionable. . . . There's a fetish with the peacefulness of the Revolution and the social media aspect. . . . Sometimes it's just simplistic. Social media was one tool that people used to promote their interests under an authoritarian regime.

The importance of face-to-face organizing

In a similar vein, Sara emphasized the importance of face-to-face organizing in any political campaign, and the value and limitations of Facebook: "Facebook is a tool that helps you to get people together [to help organize events], but it's just a tool. . . . On an organizational level, Facebook is good to organize events

. . . [and reach activists, groups and journalists]." An example that she mentioned was the march and rally planned for October 4, 2012 – the day after our interview – in front of the President's Palace to protest Article 36 in the Constitution. Organizers used Facebook, e-mail, phone calls, and text messages to encourage people to attend it. She cautioned, however, that:

> You can't use Facebook in reaching out to the grassroots, to educate, or spread information, or sensitize, and spread awareness. [The way to reach them is] getting out and talking to people – being out on the streets – organizing on the street among the people. We try to have volunteers from the same community – volunteers who live in a certain area are doing their work in their own areas . . . they say 'we live here as a neighbor; we want our streets to be safe' and ask their neighbors for their help and cooperation to make it a safe area, and that usually works.

Durar emphasized a history of organizing that predated social media use during the Revolution, and mentioned other tools people used to promote their interests:

> they went to theater [like the group that Durar had been part of], they went to political salons and gatherings, they went to . . . certain downtown coffee shops [where] people and activists are known to be . . . that are unofficial meeting places since they cannot meet as political parties, but they can meet for shisha or tea or whatever; also important are development organizations like NGOs.

She added: "When Kefaya started [in 2004] they never had a Facebook page or e-mail" but they still organized demonstrations, like the student movement in the 1970s that never had the Internet but was still very active. Music was another tool; an example she gave was Sheikh Imam – a political musician of the 1960s and 1970s whose tapes were banned for years, but the songs survived and people were singing them in Tahrir Square during the 2011 Revolution.

Technology mastery methods and confidence

Since a primary purpose of this volume is to understand the ways in which women master the use of certain technologies, it was important to discover how the interviewees learned to use social media tools, and whether their family experiences contributed to their confidence in such use. All of the women were essentially self-taught in computers and social media. When asked how she learned to use those technologies, Hasiba (the eldest of the interviewees at 70+ years of age) replied:

> At the beginning, I didn't use any of them, but after a while I found out that you have to, otherwise you'll be outside of the communication network. And it's so easy by the way.

> I attended a very short course [offered onsite by the vendor who sold the computers to the feminist organization she heads] . . . to understand the language, the basics . . . and by trial-and-error. . . . I learned it piece by piece, not one shot.

Her sophisticated understanding of which technologies best serve a particular purpose was impressive. In response to the question asking her about her level of confidence in using those technologies, she replied:

> I cannot say that I am 100% confident in using it. . . . there are things even now that I cannot do myself . . . I have to ask the help of my staff. . . . I can tell you one example: I can put a presentation on . . . , but I cannot use Prezi, which is a more interesting [presentation tool] than PowerPoint – more alive [and] more dynamic. So I have to ask one of my staff to assist me in doing a presentation using Prezi.

The next eldest interviewee, 50-year-old Bayan, stated that she learned to use social media technologies: "By courage." When her daughter began to use Facebook, I told myself: "'I must be with my daughter.'" Bayan also noticed that Twitter became important during the presidential elections in Iran (2008) and in the U.S. (2008). She indicated that when she first read about Twitter, she did not know what it was, but then: "I made a Twitter account in 2009. When activist people had accounts in Twitter, that activated me." Still, she experienced problems in using Twitter in the beginning, like not being able to find the account she created initially and having to establish a second account.

Dalal, who was 36 years old at the time of the interview, began using the Internet and computer technology in 1996 when she was an undergraduate at university, a few years after the Internet came to Egypt. In response to my question about how she learned to use ICTs, she stated: "You don't learn to use the Internet, you just click . . . and it's very easy. It's like when you learn the alphabet, then you can start to read and write." Her progression of Internet use began with e-mail, then e-groups, e-forums, and e-discussions, and ultimately expanded to Facebook and Twitter. "You just start using these things, and after that you try to use them for the causes that you're defending and working on."

Family and educational influences

Twenty-seven-year-old Sara recalled that her father purchased a computer for the family to use when she was about 14, noting that at that point middle-class and upper-middle-class families were able to purchase them, though Internet connection was dial-up and somewhat slow. She taught herself to use the computer and wasn't afraid of trying new things. Later while at university she would learn about new programs and tools from her friends, and try them.

Durar, a year younger than Sara, also grew up with a computer in the house and would use it to play games: "We had very old basic computers, but we always had a computer around . . . in our home." She stated that while her father was happy to have her use the computer, her mother started "flipping out" when a teen-aged Durar used it to chat with her friends, and wanted to know what she was doing and who she was talking to: "What really scared her was that she didn't understand what I was doing."

Like the other women interviewed, Durar had learned to use social media tools not very long before the Revolution, while at university:

> It wasn't so long ago; I started Facebook because my friends were doing Facebook . . . actually before that I used MSN chat and ICQ chat . . . at 13 or 14 because my friends at school were using it. I was basically chatting with my friends; I wasn't meeting anyone new. And then came Facebook; I kept away from Facebook for a while until everyone was there, and I felt I should use it. And then there was the blogging because I like writing and I thought that instead of writing in my agenda I'd blog.
>
> I opened a Facebook profile (about 2007 – the year before she graduated); I learned from my friends who had Facebook. . . . My friends basically told me how to use it and I figured my way around. And blogging is the same. My friends who had blogs told me how to use it. And . . . I think I'm very limited in blogging. I know you can do so much more with your blog than what I am doing with it, but I don't know everything about blogging. On Facebook, I think I'm fine. I'm not the best; other people know how to do more stuff with Facebook, but I think I'm fine.

Twenty-three-year-old Hafa, the youngest of the interviewees, was exposed to computers at an early age, and observed women using them. She recalled: "The very first computer that I ever remember seeing belonged to a . . . female cousin of my mother's. She was in marketing; she would let me play with the computer." There was a computer lab at Hafa's school from the time she was 6 or 7 years old and her teachers were always female. Her family got their own personal computer when she was about 8 years old, and both her father and mother used it for study and for work. She likened learning to use new communication technologies to learning to read, as did Dalal:

> I don't think it wasn't a matter of learning; it was just something that was around. I joined Twitter quite early – I think it was in 2008 or 2009, and at first found it very confusing, and then suddenly . . . it's just like learning reading – you can't really remember how you did it . . . it's just a day by day thing, and all my peers were using [Facebook and Twitter]. It's become the communication of choice with people in my age group . . . people don't use the telephone anymore. . . . You learn to strategize how to use it . . . and over time you learn how to use it and how to tweek it.

Hafa's initial use of those technologies was social, not political: "it was a way to stay in contact with people, and then it developed later on with time . . . [as her activities changed]. You can have a tool for a long time and change how you use it."

The younger women interviewed did not see a serious age-gap among women in the use of ICTs. Dalal stated:

> It depends on the need; people, especially women realize that the more they are involved in a public space, the more they need to use this type of technology and to learn it and to see how they can use it.

She gave the example of how the female professor who supervised her master's thesis started using computers in her late 40s. She would approach Dalal for help sending e-mail and attachments, "and now she's on Facebook and Twitter." In comparison, her mom used to be a school headmistress, but there was no need for her to use the computer in that context, so she did not do so. She believed that people who begin to use social media like Facebook do so out of curiosity: "they just go with the wave and become addicted to it." She added: ". . . definitely there is an age gap in social media use, but it is closing day by day. Because people are really getting obsessed with technology and using online tools and all these things."

As mentioned, Hafa's mother had a female cousin in marketing who used computers. Her own mother was interested in doing online marketing of the traditional handicrafts she sells. Although Durar expressed frustration at her mother's reluctance to learn to use the computer and Skype, some of her aunts were at the time of the interview "using the computer . . . because their kids are away so they have to Skype with them, others because they want to know more about the world and have Facebook and stuff." She gave the example of one aunt in particular, whom she characterized as "amazing" and a "fast learner":

> My aunt for example . . . she's turning 50 and she decided to have an e-mail and a Facebook account, and she's now the administrator of various Facebook pages; and she's turning into a Facebook activist. . . . It depends on the women . . . older women, some of them, have Twitter and Facebook. They do if for their kids, they do it to fill time; they do it to know what's happening in the world because they can't go out. [This] was impressive for me, and really broke the whole myth that old women cannot use Internet. That's not true; it depends on how receptive you are and how much you really want it.

For some of the women their parents – especially fathers – encouraged them to have careers and instilled self-confidence in them. Although Hasiba's mother was a traditional housewife, her father was a professor of History, and he "was a role model for me" in her decision to become a university professor. She was also mentored by a male professor during her undergraduate studies at the American University in Cairo: "I appreciated [him] very much; he was influential in my future."

Hafa's father had earned his master's degree in political science and worked as a diplomat in England; her mother had been a pharmacist, but had to give up her profession while serving as a diplomat's wife. Upon her return to Cairo she went back to that field, but found it boring and decided to start her own small businesses – one that promotes and sells traditional Egyptian handicrafts, and a tailor shop. Hafa was encouraged by her father in terms of technology use (her first computer was given to her by him), and her mother more generally:

> My Dad loves technology and he's always trying to find the next perfect gadget . . . my mother's not very much into it, but my mother has always been encouraging of basically anything I do; she's the one who's pushed me forward in all my endeavors – even if she doesn't always understand what I'm doing, which is most of the time. . . . She's always allowed me to advance myself in any way.

When asked what persons or experiences had the greatest influence on her choice of career, she stated:

> Not really my parents or my teachers to be frank; I've always wanted to be a writer in general . . . my mother was the one when I was younger she would tell me 'you need to find a way to make a living out of that' . . . She always thought I was a little too introverted for my own good, so she would encourage me to do things that would push me out of my shell. She didn't guide me to journalism specifically, but she . . . encouraged me to find whatever it was that made me happy . . . [Journalism] was just a way to be able to write all of the time.

Bayan's mother also played an important role in her personal development growing up. She described her father as:

> . . . a traditional man and a traditional man is always worried about girls and they want to protect us, but he didn't close a door. At the same time, my mother supported us; she was fighting for our freedom. I think if you want to make something you can do it, [despite] challenges.

Her mother had "a lot of arguments" with her father and prevailed in enrolling Bayan and her sister in a cultural center during the summers, where they learned singing, art, and broadened their horizons. That experience, and her mother, gave the girls confidence. Still, she described her entry into the field of media communication as coincidental, based on a random discovery of her talent.

Internal motivation and career choice

The career choices of the other women were self-motivated. Although Dalal's father was a journalist and political activist and her mother a high school headmistress, she decided on her own at "a very early age" that politics would be "fascinating work for me." That interest stemmed in part from watching the news as a child:

> all the time I would be glued to the TV screen, trying to understand why do these people have the authority to make decisions affecting millions of people; why are people going to war? Why do we have to fight each other?

She was especially interested in news about the 1982 Israeli invasion of Lebanon, and the first Intifada. As an 11-year-old at that time, she began to identify with "young children throwing stones at Israeli troops" and those events led to her interest in the field of Political Science:

> When I started studying Political Science at Cairo University, it automatically . . . [followed] that I would be involved in research; I wouldn't be in academia myself; I don't fit into teaching and the university image, but I would always be involved in research, so I would spend hours at the library, and at the same time I was observing people and trying to analyze the dynamics of people involved in politics, specifically women.

Dalal was also influenced by Benazir Bhutto, whom she regarded as a role model: "I remember all the time I was looking at her face when I was young and thinking she's a very beautiful lady . . . and [realized] that there was no contradiction between being a woman and being in politics." She found herself inspired by "women who decided to go into politics and really take the burden of being criticized, and all the information campaigns against them . . . I was really encouraged by all these dynamics," including wanting to find a way to influence political decisions. She added:

> We keep saying in Egypt that the Revolution of 2011 is a Revolution that had been boiling for the last 10 years. We had been working for 10 years . . . When the Revolution took, we realized that we were in the eye of the storm and we didn't realize that we had made it. . . . And I remember that when I was in Tahrir, I would have images from my childhood of people . . . in East Germany breaking the Berlin Wall with hammers. So at this point [I thought] if people can really break the wall, maybe someday my people will really seek democracy. . . . and I was fascinated by all these images.

In terms of career choice and mentoring, Sara stated: "I was quite left to decide what I wanted to do." Her father was working as a civil engineer; her mother studied political science at Cairo University; "she worked for a while and then she stayed home." Sara remembers her dad encouraging her to try math and engineering, since she was good with computers, but she herself felt that she was not strong enough in math. Instead, she gravitated toward journalism, and chose that as her undergraduate field of study. While at university she had summer internships at newspapers, magazines, and news agencies, which led her to discover that she enjoyed working on women's rights and civil society issues with NGOs.

Durar also came from an educated family; her father was an engineer and her mother a librarian, then a housewife. Yet the motivation for her career choice was internal. She stated that it was her "interest in gender" that had the greatest influence on her choice of profession and field of study. That interest was spawned by amateur acting she did with a theatre group that reenacted women's stories. She recalled: "I really enjoyed it, and I wanted to explore this further. I always was interested in women's issues," and translated publications became a source of her intellectual development in that area. She added: "I remember when I first did feminist reading; I was really taken by this."

The only Egyptian feminist writer she knew of was Nawal el Saadawi, who was characterized in the media as "crazy." Durar read el Saadawi's articles, but didn't begin reading her books until pursuing a master's degree in Gender Studies. She had also done a post-undergraduate internship with UNIFEM, and through that she was exposed to more literature on women. She recalled that during her undergraduate years:

> I was really oblivious to this; I didn't know anything about . . . gender, about feminist writing. A very small part of a course I took on political theory mentioned there was something called feminism, but that was it; I was that ignorant.

Gender-based obstacles encountered

All of the women described having encountered obstacles in life related to gender-role stereotypes. Hafa explained that while many of her female relatives had gone to law school and encouraged her, others in her extended family questioned why, as a girl, she would pursue such a degree since she would likely get married and not use it. She also experienced gender oppression in her work:

> As a journalist, some people may not want to talk to you just because you're a woman, especially if they have a certain ideology; that hasn't happened very often, but it has happened . . . and it can get to you, but you go on. And it's hard to do more dangerous assignments . . . my parents would worry . . . they would encourage me and then have fears . . . if I went down to a demonstration to cover it. There were a lot of fights [with my parents] early in my career about going here and going there and what's seemly and not seemly. But you overcome that with time; and it pushes you to work harder, and to make a point about: 'I can do the job just as well as anyone else.'

Still, those obstacles did not deter her:

> It made things difficult at the beginning . . . you don't want friction with
> people; basically my job is to go out and look for trouble, so I'm trying to
> avoid having . . . trouble in my regular day-to-day life so I can go out and
> do this job effectively. People [family members] are anxious about you all
> the time. You have to figure out ways . . . to reassure [them] that . . . I'm
> not going to risk my life or my personal safety when a lot of the time you
> may have to. You find coping mechanisms . . . in the beginning . . . it did
> sort of stand in my way, but then you make a decision – do I want to do
> what I love, or do I want to just sit at home and mope? . . . You don't do
> that; you have to just buck up and go for it, really.

Dalal gave the following reply to the question on whether she had encountered
obstacles related to gender-role stereotypes: "Yes all the time I'm asked this
question, but the point is . . . as women we have obstacles all the time." As a
young girl, she was not allowed to stay out late (at a party or with friends),
though her brother was allowed to do so:

> I've never understood this. My female friends said that parents did it for
> protection. I didn't see it as protection, but rather as an obstacle to our
> mobility. We don't need protection; we just need to be free to decide. This
> is a challenge – even for the young generation. People ask – 'why bother
> about women's rights?' I just want women to be free.

Yet the obstacles she has encountered have not deterred her: "I'm a very stub-
born person; I don't allow things to stop me; I spend days analyzing these
obstacles and how to overcome them, and then I just work around them. I
don't understand the word 'obstacle.'"

Moreover, she believes that the younger generation does not take the struggles of those who went before them for granted:

> The young generation wants to fight and totally understands that this is
> their future. . . . They are willing to set the rules of the game itself. It's true
> for both young women and young men; the 'newcomers to the movement'
> are setting the agenda. I'm learning from them . . . they are using arts,
> cinema, theater to communicate against a classist, racist society . . . And
> this is really wonderful. I'm learning from their spirit and determination
> and dedication to change.

Bayan replied that she had encountered discrimination on the basis of being a
woman, adding that her father:

> was a traditional man and a traditional man is always worried about girls
> and they want to protect us, but he didn't close a door. At the same time,

my mother supported us . . . she was fighting for our freedom. I think if you want to make something you can do it, [despite] challenges.

Her mother had "a lot of arguments" with her father, but in the end prevailed in enrolling her daughters at a cultural center, and she gave them confidence.

In Hasiba's case, it was her father who had more liberal attitudes about gender:

> Of course I have experienced discrimination, since childhood. I'll give an example. A girl has to be sweet, she has to be nice, she has to be mild . . . whereas a boy has to be aggressive, can shout, can quarrel. But I had a very cultured father [and] I don't think I experienced as much discrimination as any other Egyptian child.
>
> . . . my mother [and others] at that stage and in that generation were just wives. They didn't have ambitions; they didn't have confidence as much as professional women would have. There was a difference between my parents – as much as my father was cultured, was educated, was open to different ideas, my mother was traditional in her own way.

Durar, too, felt supported by her father, notwithstanding different expectations for girls than for boys:

> From a very young age . . . because you're a girl there are certain things that you can do and things that you can't do; and the things that you cannot do are more than the things you can do. And there are certain things you can be and certain things that you can't be, from walking on the street to what you wear, to the different treatment that your brother gets, to everything; it's just different.

When it comes to education, for example, I got very good education and my father was very happy with me getting the best education I could get. But when it comes to getting home after curfew, no, that would be difficult. . . . it took me some time to claim my space. . . . So there's this dichotomy that you're being empowered in some spheres – you have the right to work, you have the right to get a great education, but you don't have the right to live alone or to do whatever you want.

She added that the aforementioned obstacles did not hinder her:

> No, I cannot say that they have absolutely hindered me in any way. They make me very angry . . . I had a dad who really empowered me; he really encouraged me to study, and to pursue what I want. And he stood up for me when people said I cannot travel on my own or do this or do that.
>
> My mom would say that probably she wants me to not end up a housewife, and she wants me to work and stuff, but she wouldn't act like it. On the contrary, my dad might be the one who might have a conservative

discourse, but whenever there was something . . . he would support me 100%. I would say that my dad was more supportive.

Sara did not experience gender-based obstacles in her family:

> I grew up feeling independent, being told I should be independent . . . I should [be able] to earn a living, and do my own thing for myself . . . I remember at a certain point my dad wanted us, me and my sister, to travel alone so that we could take some responsibility being totally on our own and knowing what to do and how to think. I wouldn't say everyone around us was the same. My dad had this vision and it was good and helpful because it helped us develop a lot.

Sometimes she would be told not to be out late since it wasn't safe for women, but stated: "I think this happens to everyone, but it didn't affect my life." She would later discover, though, why her parents were concerned for her safety: "The problem was going out on the street and being independent as my parents wanted me to be and seeing how the street reacts to women in a certain way, which is sexual harassment."

In college when she was working summer journalism internships:

> I would be shocked by the amount of sexual harassment I would encounter in one day, and that was basically telling me 'You're not welcome here; you shouldn't be on that street. . . . You have to struggle; it didn't stop me from doing anything but I had to face it every day.

She added that sexual harassment is not unique to Egyptian society, asserting: "Sexism is everywhere." Indeed, sexual harassment's prevalence in the U.S. and elsewhere is well known (Chatterjee 2018; Williams 2017).

Gender-role stereotypes did not limit girls in the educational sphere. All courses of study have been open to women and girls, though Bayan cautioned that while that has been the case "until now, we don't know what will happen" reflecting her fear that Egypt could go the way of Iran in limiting women from technical courses if conservative Islamic law becomes embedded in the Constitution.

Identification with feminism

Every one of the women interviewed identified strongly with the term "feminist," and did not regard it as a label imposed from outside Egypt, especially by the West. Sara stated:

> I say I'm a feminist, and I don't mind the label . . . a lot of girls here are afraid of saying they're feminist cause that would prevent the boys from talking to them, but I don't care.

> I don't have this vision that it's a . . . term being forced on us, because it's very fluid and diverse; and it has a lot of versions and perspectives; there's not only one feminism; it's like Islam – there's not only one version of Islam. . . . But for the basic core values of feminism, I fully agree with them. There's been a history of Egyptian feminists struggling since the 18th century, so I don't think it's a new thing being imposed on us, and I'm not afraid to say I'm a feminist.

Durar said that she identifies as a feminist "very confidently. I was hesitant for a while that I would identify with this term." She further explained what feminism means to her:

> It's not only fighting for women or standing up for the rights of women . . . it's about really pushing forward for an egalitarian society in whatever way it could be. It's just about egalitarianism, whether it be economic or social, in whatever context.

Hasiba commented on the reincarnation of the Egyptian Feminist Union, and the importance of retaining the name used by its founder, Hoda Sharaawi, "because it is sometimes refreshing to go with tradition" and the term is accepted. She is aware that "some people don't like the term feminist, but I use it and I have no hesitation in using it." Being a feminist to her simply meant: "I'm concerned about women's issues and I'm ready to work on them – that's it."

Bayan explained the nuances of feminist identity in Egypt following the 2011 Revolution: "Here in Egypt, they try to destroy the terminology . . . sometimes when you talk about quotas in the Parliament for women, the public doesn't support it." She described the reluctance of the newly formed Constitution party headed by Dr. Mohammad ElBaradei to focus specifically on women's issues, and the difficulty of having a separate women's party. Women like her who were inclined to create a political group prioritizing women's issues opted instead in the spirit of solidarity to work on national matters. She added:

> But after the Presidential campaign I took a decision I must return [to] working on women's issues, because before the Revolution and . . . through the Revolution there was a lot of violence against women from the Army and the police. . . . And in the new Constitution they want the woman to be married at 9 years old. . . . The violence in the streets against women has become a sin, and we can't talk about society and personal attitudes without talking about what creates it – it's created by the police, and the army, and now the Constitution.

Dalal also discussed feminism in the context of political life in the post-Mubarak era:

70% of the people I meet understand that feminism is about women's rights . . . Feminism for me is about choice, and about freedom and about solidarity – it's all these values, and it's also about mainstreaming women's issues into health, education, and so on. We try our best, especially after the Revolution, to have a space open for us to discuss these issues in political parties and within civil society and social movements. This is helping us to spread feminist values.

She revealed that not all her friends understand the term "feminist," and one public perception is that it refers to man-hating, a misrepresentation that exists in the U.S. as well.

Hafa articulated her support for feminism in this way:

I'm very much a very proud feminist; I identify very deeply with the word. . . . I'm generally not a fan of labeling, but it's one of the few labels I accept; it's very important as a way of life and as a set of beliefs . . . it's a very big component of the core of my life, really, being feminist.

She defined feminism in the following humanistic manner:

Basically it's a refusal to let inequality slide . . . and it means ensuring that people see people as humans – nothing else . . . that's my ultimate dream. There's no longer any identification as male or female or as this race or that race; we just identify each other as humans. Feminism is all about that; that's what it means to me.

Benefits and challenges of political activism

A number of the women emphasized that their involvement in political work on behalf of women's equality, social justice, and democracy increased their confidence more generally, though in some cases with costs as well. Sara indicated that:

it does help you explore more new areas, talk to more people . . . but it also frustrates a lot and puts a lot of strain and stress on you, because it's a struggle and a continuous one – trying to be heard. It's not easy. . . . It might be empowering, but it's also stressful.

Bayan also described the stress borne of struggle: "It was a very hard experience. I saw the Revolution . . . it has caused a lot of conflict between people. . . . the atmosphere is not good here. Very stressful."

Durar characterized the benefits and disadvantages of her activism as follows:

The benefits would be exposure. . . . We put a lot of things on the Revolution, I don't think it's only the Revolution; it's also we were growing up within it; I was younger and now I'm older; I was less educated and I'm now more educated. So it's not only my participation in the Revolution; it's a natural process of growing up. It did speed things because you see things and experience things that you would not necessarily experience.

An additional benefit was her discovery while doing the research for her master's thesis. Contrary to her initial assumption that the Revolution empowered women, what she found instead was that:

> women empowered and fueled the Revolution in so many ways. . . . A friend of mine said – women have always been fighting for space; they just came and brought this energy, and it just blew up the fight. And it made me less scared of many things; I'm far less scared of the government, I'm far less scared of men harassing me in protests . . . my mom still tells me don't go; I'm far less scared than my mom. . . . I'm just far less scared of things, and I'm more exposed. I wasn't in a political party before and now I'm in a political party. My network has expanded locally, regionally and internationally.

She acknowledged disadvantages as well:

> It's the ups and downs – you have so many dreams and then suddenly you're very disappointed and you have to learn to live with that. It's so crazy, and things are changing at an extremely fast pace, and you have to keep up with various struggles at the same time.

Still, some of the women described finding comfort in the friendships they built or had. Hafa stated:

> It's good to be with like-minded people and to have your ideas vindicated, so to speak. You form these really great friendships as well; you create a community of people – a support system is very necessary to survive every day in day-to-day Egypt because sometimes you get very down about how difficult everything is and how huge the fight is to get the things you believe are right [accomplished]. But having all these people around you is very reassuring.

Dalal also described the ways in which her friends have helped to ground and recharge her:

> On a very personal level, sometimes I realize that I don't have time for myself . . . sometimes I get disoriented and need to have some space for myself. But I'm one of those lucky persons who have close friends who let me realize what is important and put me once again in a laughing and high-spirit mood.

In an effort to be light hearted, she revealed that she and her friends might discuss Turkish drama on TV instead of politics, adding: "Sometimes I forget about my well-being. I'm getting more attached to my mobile and Internet and I'm getting obsessed about getting news through Facebook and Twitter . . . I'm trying to de-attach myself from my Blackberry."

Post-Revolution concerns

The women activists also voiced concern about women's rights in a post-Revolution Egypt. Durar recalled with dismay that when women were attacked verbally at an International Women's Day rally on Tahrir Square "On the 8th of March 2011 less than 1 month after the Revolution" it was a rude awakening that:

> basically women's rights have no place in the national struggle and this is the first thing you realize. And then comes political Islam and then comes the death of so many people. So [this affects] all the dreams you have for the country, all the dreams you have for yourself, all the dreams for your own political career . . . it's very hard to get over this.

Bayan offered important information about the way that women fared politically in the aftermath of the Revolution:

> After the Revolution there has been a marginalizing of women because those who are in the decision making positions are not very much in favor of equality of women. So we are trying hard, we are putting so much effort, trying to raise our voices and trying to get ourselves inside the circle of decision making, trying to be partners and continuing to be partners. Because we were partners during the Revolution, and we shouldn't be kicked out after the Revolution. Right now we are working – trying hard, to contribute to the drafting of the Constitution. This is very important. . . . If the Constitution comes out without admitting equality between the sexes, it will be very difficult for us afterwards to exercise our rights.

Bayan and other interviewees also expressed dismay about the overt sexual harassment and physical assault experienced by women during and after the Revolution, persisting past the national presidential election of Mohamed Morsi in May, 2012 (El Deeb 2012a; Ezzat 2012). One of the most egregious examples was the so-called virginity tests administered to seven women arrested during the demonstrations in Tahrir Square, "allegedly as a way to ward off claims of sexual harassment while in police custody. The patriarchal rationale being . . . virgins cannot claim rape, and only rape constitutes sexual harassment and/or assault" (amadorsquare 2012). Another was the public beating and stripping of a young woman, who came to be known as the "blue bra girl" by soldiers in Tahrir on Friday, December 18, an event caught on film and broadcast across the world (Amaria 2011). Those incidents rallied over 10,000 women and men to march in Cairo in December, 2011, shouting "Women are a red line that cannot be crossed" (Morrow and al-Omrani 2012; El-Naggar 2011).

Men, too, were subjected to violence at the hands of police and the SCAF. Twenty-eight-year-old Khaled Saeed (also spelled "Said") of Alexandria became an early martyr whose death on June 6, 2010, fueled the Egyptian Revolution. Although the police alleged that Saeed had died of drug-related causes, witnesses reported having seen him being brutally beaten to death by detectives outside of an Internet café (*Egypt Independent* 2010; Schemm 2010). He had apparently come into possession of a video implicating the detectives dividing drugs and cash seized in a drug raid, and relatives – suspecting foul play in his death – bribed a morgue guard to take a picture of his battered, disfigured corpse (Londono 2011). The photo went viral on the Internet, and Wael Ghonim, a young Egyptian marketing executive working for Google in Dubai, was moved to establish a Facebook page entitled: *We Are All Khaled Said*, for which he himself would be detained for 12 days by Egyptian security forces while in Cairo during the Egyptian Revolution (England and Saleh 2011; Giglio 2011). He subsequently left Egypt for the United States, and co-founded a discussion platform called "Parlio" – designed to promote civil and intelligent online conversations through informed reading and pledges of civility (Buhr 2016, 1).

The status of women in post-Revolutionary Egypt

Feminists involved in the Egyptian Revolution of 2011 resulting in the over-throw of President Hosni Mubarak discovered that their political contributions did not confer upon them true equality (Egyptian Center for Women's Rights 2014). That was reminiscent of what had occurred in 20th-century struggles for gender equality, as when Doria Shafik engaged in a hunger strike in 1954 to demand women's inclusion in a new constitutional committee and was instead ultimately placed under house arrest by the Nasser government (Zakhary 2016; Ragai and Ragai 2014; Nelson 1996).

Not only was the post-Revolution Constitution written without assuring full equality for women, but some women complained during the December 15, 2012 constitutional referendum of efforts to suppress their votes (El Deeb 2012b). Dr. Hoda Badran, chair of the reincarnated Egyptian Feminist Union, noted that after the 2011 Revolution and the transition to a ruling military council, women continued to be underrepresented in the Parliament and in governates (Gray 2012). Although a quota was put in place in 2010 to increase women's presence in the People's Assembly (following earlier similar actions), in general it had a limited effect and did not survive:

> The women's nomination quota was revoked in the new electoral laws. Currently, the election law says that each PR [proportional representation] list must include one female candidate, but does not oblige the party to put a woman in winnable positions on the list. As outlined earlier, the 332 PA

[People's Assembly] seats elected through proportionality by party lists are elected in districts that range from four to 12 members in size. However, less than a quarter of the 46 districts are as large as 10 or 12 members and the vast majority of districts are four to eight members. Based on current polling trends, it is likely that parties will win a maximum of one or two seats per district. It is possible that a single party could win three (or four) seats in the 11 districts of 10 or 12 members in size. This means that in smaller districts if a female candidate is not in the top two positions, she will stand little to no chance of winning.

If every party were to place a female candidate in the second spot on their list, it is likely that such a candidate would secure a small, but significant portion of seats. . . . However, if larger parties place a female candidate third on the list, instead of second, then the number of female candidates that win will drop dramatically. It remains to be seen how well female candidates fare running for IC [individual candidacy] seats. Past election data indicates that female candidates not running for a PR seat have fared poorly.

(International Foundation for Electoral Systems 2011, 10)

A 2011 conference organized in part by the Alliance of Arab Women put forth a "roadmap to arrive at democracy and reach the goals of the Revolution" and along with the Egyptian Feminist Union sought to "reach women outside of Cairo and try to make them involved in the political process" by encouraging "women to vote" and "be candidates" and "providing poor women ID cards" so they could vote (Raghall 2013, 2).

There was also concern about whether language in the new Egyptian Constitution would assure women's legal equality with men, especially when there was an agreement among members of the Constituent Assembly drafting the Constitution "to repeal Article 68" which recognized "equality between men and women" (Dalsh 2012, 1; Abdoun and Caspani 2012). Legal scholar Marwa Sharafeldin, co-founder of the Network for Women's Rights Organizations, wrote in March, 2012:

If the new constitution only recognizes the full 'citizenship' of Egyptian men because they are the strong able citizens of this country, and discriminates against women, denying them full citizenship rights under a pretext of being subordinate 'harem', it would be a catastrophe. Our previous 1971 constitution seemed to do just that. . . .

Additionally, the claim that the citizenship of women is inherently limited by Sharia often ignores both the early history of Islam . . . as well as overlooking the changing historical contexts and the developments in scholarships in the interpretation of the Quran. Scholars have conducted much rigorous research across the world based on Islamic scripture and jurisprudence. New, unapologetic grounds have been broken by the

production of this knowledge and laws, where women and men are considered equal citizens despite their biological differences. . . .

We should not leave it to others to tell us what our own religion tells us
. . . to do, especially if these others are politicians. We should participate,
like other Muslims across the world, in the ever-evolving and dynamic
creation of this religious knowledge, building on our exceptional heritage
of jurisprudence.

(Sharafeldin 2012, 1–2)

Another Egyptian feminist observed at that time:

The draft represents a vision that sees women's perfect place as in the
house as a wife and a mother while the state could help her if she becomes
a widow or gets a divorce. It refers in article (10) to the role of the State
and the society in maintaining the authentic character of the Egyptian
family, and how they should work on its cohesion, stability and protection
and of its traditions and moral values. . . .

I could go on analyzing every article and their contradictions, but what I
care about most is the philosophy embedded in the draft. A philosophy that
praises conservative social norms for women, children, youth, ethnic and
religious minorities; a draft that introduces a political system that would be
hard to describe as democratic in which the military keeps its special political and economic gains; and an economic system that adopts many of the
guidelines of neo-liberalism.

(Abdelaal 2012, 2)

According to one account, "Women comprised fewer than seven percent of the
Constituent Assembly that drafted the new [2012] constitution" (Patterson
2013, 3). Feminists wanted a "quota system" in the Constitution to assure "a
reasonable number of women in parliament" since the "first parliament since
Mubarak's ouster had a sweeping majority of Muslim Brotherhood and Salafist
MPs, with only 10 female deputies all hailing from the same party" and no
women winning individual seats (El Tahawy 2013, 1). An observer wrote:

With Islamists . . . dominating the parliament, the presidency, and the
assembly that was drafting the post-revolution constitution, and with liberals and secularists struggling against the establishment of a religious
state, a heated debate ensued. . . . about the link between Islam-related
articles and Egyptian identity.

(Farid 2013, 2)

Of particular concern to feminists was Article 10 entitled "Family as the basis
of society" which read: "Family is the basis of society and is based on religion,
morality, and patriotism. The state protects its cohesion and stability, and the
consolidation of its values" (Constitute Project 2016, 13).

There are different views on how women faired in the 2012 Constitution. With reference to an earlier article by El-Ghobashy (2005, 382), McLarney (2016) wrote that in the 2012 Constitution, "the liberal language of women's rights and equality converged with Islamist political aims" as the "Morsi government adapted the liberal language of women's rights, drawing simultaneously on a long history of constitutional language as well as a long history of Islamic rhetoric about women's rights in Islam" (2016, 2). Noting that "the U.S. constitution has no mention of equality anywhere between anyone" McLarney nonetheless pointed to limitations of the Egyptian constitution:

> Gender inequality remained encoded in the personal status laws with regards to witnessing, polygamy, and divorce. But the liberal language of the 2012 constitution sublimated these inequalities (in typical liberal fashion) underneath euphoric celebrations of newfound political liberties, pluralism, democracy, and freedom (mentioned no fewer than eight times in the preamble alone). The language clearly rankled activists, who along with feminists, critiqued this liberalism's dissimulations and hypocrisies, along with its dualism and paradoxes.
>
> *(loc. cit.)*

During that early post-revolutionary period, the head of a United Nations working group criticized not only the language of the Constitution as failing to guarantee women's rights, but also the underrepresentation of women in the process of drafting the document (Gulhane 2012).

The 2012 Constitution was suspended following the ouster of President Morsi, and a "new" Constitution was adopted in 2014, which "de jure" was built upon the prior one, but with 42 new articles and a new preamble (Megahed 2014, 1). Women organized during the referendum, just as they had been doing prior to and during the Revolution. A member of the International Alliance of Women observed that during the referendum of January 14–15, 2014 "it was obvious that women overwhelmed men in many of the polling stations" and the new Constitution was approved "with significant participation by women" *(loc. cit.)*.

Ultimately, the Constitution was written with women's rights explicitly affirmed. Article 11, entitled "The place of women, motherhood, and childhood", reads:

> The state commits to achieving equality between women and men in all civil, political, economic, social, and cultural rights in accordance with the provisions of this Constitution.
>
> The state commits to taking the necessary measures to ensure appropriaterepresentation of women in the houses of parliament, in the manner specified by law.
>
> It grants women the right to hold public posts and high management posts in thestate, and to appointment in judicial bodies and entities without discrimination.

> The state commits to the protection of women against all forms of vio-
> lence, and ensures women empowerment to reconcile the duties of a
> woman toward her family and her work requirements.
>
> The state ensures care and protection and care for motherhood and child-
> hood, and for breadwinning, and elderly women, and women most in need.
>
> (Constitute Project 2016, 13)

The new Constitution was said to have offered greater freedom of speech for individuals and the press, but with restrictions that worried activists (Al-Tawy 2013). Another concern was that while Article 93 committed the state to comply with "international agreements, conventions, and charters" that it had ratified, such as the Convention on the Elimination of All Forms of Discrimination against Women (CEDAW), it placed them on "an equal footing with Egyptian law rather than granting the former a higher status" allowing the possibility for subsequent legislative amendment (Agora Moderator 2014, 3). Another critical analysis of the Constitution was written by a European Parliament policy ana-lyst, who apart from acknowledging stronger gender-equality provisions "in aspirational terms" raised concerns about "the protected role of the military" which "operates outside civilian oversight" and "plays a significant role in civi-lian matters", lack of a formula on "other forms of ill-treatment" in Article 52 on torture, "excessively strong presidency, upsetting the separation and balance of powers," and the process of drafting and voting for the Constitution which "did not provide an environment for a genuine choice" or "conform with international obligations for political participation" (Meyer-Resende 2014, 8, 11, 15).

A similar procedural criticism was made of the method by which President Abdel Fatah al-Sisi, who initially assumed his position via military coup, was elected, with one U.S. observer pointing to "a repressive political environment" that made a "genuinely democratic presidential election impossible" (Kirkpa-trick 2014, 1). Following the ratification of the Constitution and the presidential election, the parliamentary elections of 2015 were the final step in the post-Mubarak, post-Morsi government transition. An analysis of those latter elec-tions by the Egyptian Center for Women's Rights found that despite the diffi-culty female parliamentary candidates faced given "tribal fanaticism, the control of political money . . . , financial inability, and most of the parties nominating a limited number of women on their electoral lists and none on the individual seats" women secured 19 individual seats and 89 seats overall for a total of 14.7 per cent – the largest percentage ever in Egyptian parliamentary history (Abol-Komsan 2016, 16).

However, the post-Revolution fears expressed by interviewees about women's status, government control, and free speech continue to be validated by more recent government actions. According to Human Rights Watch, Mozn Hassan, Founder and Executive Director of Nazra for Feminist Studies, was prevented by the internal ministry of passport control officers from boarding a plane at Cairo airport on June 27, 2016 to attend a regional meeting of women human rights defenders (Human Rights Watch 2016, 1). The same report noted an "increasing number" of court-approved "travel bans and asset freezes against human rights defenders and political activists" in the first half of 2016, following investigations of non-governmental organizations receiving foreign funding "without government authorization" (Ibid.). The government's actions against Nazra and other human rights-oriented NGOs were part of "Case 173" in which leaders like Hassan were summoned to court and assets of their organizations were frozen (Mohie 2016). Despite domestic and international protest (Nazra 2016; Global Fund for Women 2016), Nazra – founded in 2007 – had its assets frozen which forced Nazra to close its office in March, 2018 (Euromed Rights 2018).

Print and online journalists have also been targeted by the government. Mostafa (2015) wrote after the Sisi government had assumed power that the Egyptian media outlets were supporting the "military regime" and were "mobilizing their means and resources not only to *erase* the revolution from Egypt's history, but to distort it, to make it look as a *diversion*, a *disruption*, and a *conspiracy*" (121–122). Liliane Daoud, a British-Lebanese journalist who hosted the talk show "The Full Picture" which was critical of the Egyptian president Sisi's government and the Morsi government, was deported to Beirut from her home in Cairo on June 28, 2016, and was forced to leave her Egyptian-born young daughter behind (Associated Press in Beirut 2016, 1). In May 2016, three leaders of the Egyptian journalists association were detained and charged with harboring fugitives, publishing false news, and inciting protests (Habib 2016, 1).

The Committee to Protect Journalists reported that as of June 1, 2015, at least 19 Egyptian journalists were being held in prison for their reporting, "the highest in the country since CPJ began recording data on imprisoned journalists in 1990" (Committee to Protect Journalists 2015, 1). According to CPJ, government arrests of journalists in Egypt has continued; three journalists were taken into police custody and jailed in 2018 "after reporting on government opposition figures and irregularities in Egypt's recent presidential election" (Committee to Protect Journalists 2018, 1), and in 2017, Egypt ranked just behind Turkey and China as the world's top jailers of journalists in 2017 with those three countries accounting for 51 per cent of a record 262 journalists jailed worldwide (Beiser 2017, 1).

Conclusion

The Egyptian feminists described in this chapter used and continue to use a variety of social media tools in their work for democracy and equality, enabling local organization and global amplification of their messages. Their mastery of those technologies had occurred prior to the Revolution for the most part, primarily through informal learning in the form of trial and error. In some cases family members, friends, and staff provided access to technologies, knowledge, and/or encouragement, but the motivation to learn was internal. All of the women were well educated, with undergraduate and in some cases graduate degrees. While each of them encountered gender-based obstacles of one sort or another in their lives or careers, they were not deterred by them.

They used social media technologies in nuanced ways, recognizing benefits and limitations of specific tools and adapting them to particular needs like organizing political events or publicizing activities in real time. They also relied upon traditional methods of face-to-face organizing, as described by Sara in this chapter, and consistent with Mojab's earlier findings (Mojab 2001) pertaining to Kurdish women. The recognition of social media as a mere tool in political organizing has since been emphasized by Tufekci (2017) in relation to the 2012 Gezi Park and Taksim Square protests in Istanbul.

The women's passion for social justice, gender equality, democratic governance, and for freedom from political and sexual harassment fueled their courageous political work during the 2011 Revolution, and also prior to and after that event. By using the "digital megaphone" of social media, they were able to connect with Egyptian women and men and global supporters to advance their cause, and continue to do so in the face of an uncertain future.

3 Sound sisters

Engineering women's music

The 1960s and 1970s produced youth culture subgroups striving for independence from corporate and government domination. In some countries, their aspirations were to be free of colonial rule, but in the U.S., the struggle was largely directed against big business, pollution, the Vietnam War and other aspects of U.S. foreign policy, and entrenched systems of racism and sexism.

Emerging from that rebellious spirit was a series of self-help movements and activities. Food cooperatives (co-ops) flourished in cities and towns adjacent to college campuses. Bicycle and auto repair co-ops formed in cities like Cambridge, Massachusetts at which people could learn to do their own vehicle maintenance and use tools onsite for a nominal fee. Women took health matters and self-exams into their own hands – literally, aided by the classic book: *Our Bodies Ourselves* (Boston Women's Health Book Collective 1971). Women sought to demystify technology for reasons of control, confidence, and cost savings, as evidenced by the introduction to a "Dell Purse Book" reprinted in 1976 (based on a 1974 text) and sold in grocery stores to a mainstream consumer market:

> More than 44 per cent of the drivers' licenses issued today in the United States are issued to women. The first known woman driver in the country, Genevra Delphine Mudge, drove in New York in 1898—and raced the following year. Isn't it time we learned about the cars we drive? It's really not difficult to understand your car, what makes it run, how to give it loving care. And you don't need brawn, either, just normal intelligence. If you can run a sewing machine, knit, change a typewriter ribbon or use a pressure cooker, consider yourself mechanically inclined.
>
> Why should you take the time and trouble to know about cars? Pride, for one thing. If you've ever been annoyed at finding yourself at the mercy of mechanics who speak trippingly of camshafts and exhaust manifolds, you'll be able at last to look these fellows in the eye and discuss technicalities. And there's money. If you've been staggered by exorbitant repair bills and have harbored dark thoughts about the true costs of labor and parts, you'll be able to judge a mechanic's work and honesty yourself.
>
> (de Roulf 1976, 2)

While that booklet encouraged women to be knowledgeable about their cars to avoid being cheated or demeaned by male mechanics or to have the ability to perform certain maintenance and repair tasks themselves, femininity was stressed. That was evident not only in the discussion of women's mechanical aptitude derived from typing, sewing, and cooking, but also by the booklet's illustration of a woman changing a tire – in a dress and heels! In contrast, articles in feminist publications urged technological self-reliance without apology or concern for gender stereotypes, and sought to demystify the inner workings of equipment through detailed textual explanations and diagrams. Such was the case with an article appearing in a 1977 issue of the journal *Lesbian Connection* submitted by an unnamed reader from Springfield, Oregon entitled: *How to Fix a Toilet* – complete with a hand-sketched illustration.

Another traditionally male-dominated domain in which women demonstrated technological competence was the music industry. The presence of female musicians on stage beyond the realm of classical music performance certainly predated the era of second-wave feminism. Two African-American "all-girl" jazz bands of the 1940s – the "Darlings of Rhythm" and the "International Sweethearts of Rhythm" (the latter group including performers of Asian-American, Hispanic, and Native-American backgrounds) – not only exhibited great musical proficiency (Tucker 1998), but also performed songs that crossed gender roles and boundaries, such as the Sweethearts' 1946 recording *Digging Dyke* (Stash Records 1978).

In the blues and country genres, women like Elizabeth Cotten and Maybelle Carter had also broken ground prior to the emergence of the women's movement with their masterful and unique styles of playing acoustic guitar. They became staples of the folk music scene in the 1950s and beyond.[1]

The women's music movement

While women performers and songwriters earned some recognition and benefitted economically from their talents (Davis 1998), their fame and fortune paled in comparison to that of their male contemporaries – especially white men. That was indeed the case when Elvis Presley's 1956 recording of *Hound Dog* "eclipsed" a 1953 version sung by Willie Mae "Big Mama" Thornton (Gaar 1992, 1–2). The advent of the women's music movement (also known as "womyn's music") in the 1970s was in part a reaction to a male-identified and corporate-controlled music industry of many genres – particularly rock and roll, folk, and blues and jazz. It is reported that the first National Women's Music Festival (NWMF) held in Champaign-Urbana, Illinois in 1974 was organized after women had been prevented from performing at a folk festival in that state (Tilchen 1984, 7). In a similar vein, British singer Frankie Armstrong was simultaneously booed by men and cheered by women attending the 1973 Fox Hollow Folk Festival following her forceful *a cappella* rendition of Peggy Seeger's feminist song: *I'm Gonna be an Engineer* (Seeger 1971).[2]

Feminist songwriters and lyrics

Women's music was characterized by powerful lyrics written by the performers from the perspective of women's lived experiences. Certain songs became feminist and lesbian- movement anthems that both inspired and reflected women's political activism and the woman-identified culture that blossomed in the 1970s. Examples were Maxine Feldman's *Angry Athis* (Feldman 1969), Alix Dobkin's *Talking Lesbian* (Dobkin 1973), Margie Adam's *Would You Like to Tapdance on the Moon?* (Adam 1973), Meg Christian's *Ode to a Gym Teacher* (Christian 1974), Cris Williamson's *Song of the Soul* (Williamson 1975), Bernice Johnson Reagon's *Joanne Little* (Reagon 1976), Ginni Clemmens' *Solid Ground* (Clemmens 1976), Mary Watkins' *Don't Pray for Me* (Watkins 1977), Teresa Trull's *Woman-Loving Women* (Trull 1977a), Holly Near's *Fight Back* (Near 1978), and Kristin Lems' *Mammary Glands* and *Ballad of the ERA* (Lems 1978; 1982, 2007). There were also many lesser-known songwriter-performing artists with loyal regional followings – Casse Culver; Jeanne Mackey and Mary Trevor (later adding Kris Koth to form the group "Lifeline"); Cathy Winter and Betsy Rose; and members of groups like Lilith and the Reel World String Band, to name just a few.

Some popular feminist songwriter-performers also revived the lyrics of an earlier generation of blues singers and songwriters, as was the case with Teresa Trull's recording of the 1920s song *Prove It On Me Blues* written by Gertrude "Ma" Rainey (Trull 1977b), and the feminist group Alive's recording of the 1924 Ida Cox song *Wild Women Don't Get the Blues* (Alive 1981). Additionally, the women's movement benefitted from feminist-oriented folk songs of contemporary writers and singers like Hazel Dickens and Malvina Reynolds.

All-female musicians and singers

A second characteristic of the women's music movement was mastery of a wide variety of musical instruments, allowing performers to appear on stage as members of all-female bands or with other women as back-up musicians. The skill and versatility of musicians like guitarists June Millington and Nancy Vogl, pianist Mary Watkins, percussionists Linda Tillery and Carolyn Brandy, and fiddlers Barbara Higbie and Robin Flower enhanced the music of the groups with which they performed. Yet as mentioned earlier, "all-girl" bands populated with gifted musicians had existed before the 1970s.

Women-controlled production and distribution

What really distinguished the women's music movement was the expansion of women's presence beyond writing and performing music for and about women, into all aspects of music production and distribution. Women created their own record companies like Redwood, Pleiades, and Olivia – named after the lesbian protagonist in a 1949 novel by British writer Dorothy Bussey (Pepper 2008).

Feminists also created national-international distribution companies to promote and sell the recordings of women, namely Ladyslipper Music founded by Laurie Fuchs in Durham, North Carolina, and Goldenrod Music founded by Terry Grant in Lansing, Michigan.

Women's music was disseminated not only through recordings, but also at locally produced concerts and women's music festivals like the aforementioned National Women's Music Festival (Armstrong 1989; Garlington 1977), the Michigan Womyn's Music Festival (Sandstrom 2002; Morris 1998), the Sisterfire festival (McKnight 1987), and a host of regional women's music festivals. Whether their performances were live or recorded in studios, women musicians depended on the expertise of audio engineers – a technical profession that had been a male domain. Boden Sandstrom – an early audio engineer in the women's music circuit – explained why women were interested in learning the technical aspects of music production:

> Within the framework of the creation and marketing of Women's Music, women began to fully control the technical and artistic components of the circuit, including audio engineering, lighting design and operation, management, and promotion. Their intention was not only to achieve economic control of their lives but also to create a women-identified environment that would be safe and comfortable for all participants. An entire Women's Music network developed which was the economic backbone of performers and women's businesses.
>
> (Sandstrom 2002, 96)

Women engineering music

Audio engineers in this study

Audio engineering is where women displayed their greatest mastery of non-traditional, complex technology, and for that reason it is the focus of this chapter. Data were collected by means of telephone (and one face-to-face) interviews in 2008–10 with five of the pioneer audio engineers in women's music and beyond. Three of the women interviewed worked during that period as audio engineers on a full-time basis – Karen Kane, "Liz" (pseudonym), and Leslie Ann Jones. One of the the interviewees – "Christine" (pseudonym) practiced audio engineering part-time alongside an academic career. Yet another – Boden Sandstrom pursued an academic career after having spent decades in the industry. Although all of the women are similar in age, race, and class, they were based in very different parts of the country – the East and West coasts, the Midwest, and the Southeast.

At the time of the interviews, each of the five women had worked in the industry for more than 30 years, and their professional accomplishments were stellar. Boden Sandstrom had co-founded with Casse Culver and served as chief engineer for Woman Sound – an all-woman company that did technical production and audio engineering for major national rallies sponsored by the National Organization for

Women, and LGBTQ rights organizations women's music festivals, and events at Robert F. Kennedy Stadium, the Smithsonian Institution, the Kennedy Center, and other venues. When interviewed for this study she had earned a Ph.D. in Ethnomusicology from the University of Maryland, and was serving as a lecturer for that program and technical director for the School of Music – positions she would hold until her retirement in 2013 (Rosendahl 2013). Liz was the regular audio engineer for a major national women's music festival, a lesbian-run travel company, and indoor and outdoor concert venues featuring classical, popular, jazz, and world music. Christine was a frequent audio engineer at a folk and acoustic music club, and for another popular national women's music festival. Karen Kane had earned 2 diplomas in audio engineering from the Recording Institute of America, worked with artists like Tracy Chapman, Bare Naked Ladies, Harry Manx, Kay Gardner and Mary Watkins, and produced albums nominated for Canada's prestigious Juno award. She would go on to be named "Best Producer of the Year" (2013) in North Carolina, and serve as the recording technology instructor at the University of North Carolina, Wilmington since 2012 (Kane 2017). Leslie Ann Jones was and remains Director of Music and Scoring at a major sound studio in California, and has been a three-time Grammy Award winner; she broke barriers as the first female audio engineer to have been hired by ABC Records in 1975 (Pettinato 2013–2017; Hamlin 2007).

Nature of the work and technology used

The term used most commonly by the interviewees to describe their profession was "audio engineer." They explained that there are two primary categories of audio engineers: 1) recording and mixing engineers who work with the artists in a recording studio; and 2) live audio engineers who travel to concert halls and outdoor festivals to mix the sound onsite. Although most of the interviewees have worked in both domains, two of them – Karen Kane and Leslie Ann Jones – work primarily in recording studios, and the other three – Liz, Boden Sandstrom, and Christine – are live audio engineers. Karen explained that there is a further division of labor in live audio engineering: the "house mixer" mixes sound for the audience, while the "monitor engineer" mixes monitors for the musicians onstage. She added that those roles may be fulfilled by a single person. Another live audio engineer – Boden – described how her work branched out to include technical production at major national rallies and festivals, including not only mixing the sound, but also organizing video production, the large-screen TVs, satellite production, and even contracting for the porta-potties, trash removal, security, tents, chairs, tables, and catering.

Liz explained how the process of live audio engineering occurs:

> The set up begins with audio engineer or technician setting the microphones onstage, and then the signal from those microphones goes through cabling to the sound board, where it's mixed from many inputs into a couple of outputs. So the sound comes into the console, and then it gets mixed and/or equalized – which is a fancy word for tone control, and lows

and highs being added or subtracted, or compression being added, or effects put on it. And then it comes out to the loud speakers.

Christine, who also does live audio engineering, added that:

> Essentially what you're doing is getting a certain level of signal of the sound so that it actually can be heard without being too loud or too soft, for each individual thing that you're trying to work with, and then putting it all together by balancing the relative levels and the kind of sounds; you can manipulate the sounds by changing the EQ [equalization] – . . . boosting treble, or boosting bass, or cutting this and that, sometimes to your own taste, sometimes to the taste of the artist.

The technology used for both live and recorded audio engineering is similar. Liz named the key pieces of equipment used in live audio engineering:

> [There is a] mixing board [also known as a] sound board, [or] console; the English call it a 'desk'. And that's the heart of the system, and there's a loudspeaker system as well, whether it's speakers that are called high-pack, sitting on top of sub-woofers, or there's the newest speaker system technology, which is a line array.

The console/mixing board/sound board can be analog or digital, as Christine explained:

> You divide the frequency spectrum up into some discreet number of parts; 31-band is typical in an analog system, though there is also 15-band; typically the mixer has up to 4 different channels of EQ, so it's not treble and bass, but it's high frequencies, and high-bit frequencies and low-bit frequencies, and low frequencies. In the digital world, of course, you can dial in whatever particular frequency you want to dial in, and boost or cut that.

Leslie articulated the importance of having a working knowledge of analog mixing in a digital age:

> I think that to work with analog equipment, you need to understand some of the principles behind what it is you're doing more than you would if you were using digital equipment with sort of a representation of what that is. [I approach technology use] from a musical standpoint and have learned what I've needed to know to be able to use the equipment that I have to use, but I do care how it sounds when I plug something into it . . . and how the controls manipulate sound. I think that might be lost on people who don't have the advantage of having some analog background.

Karen described additional pieces of equipment used in live and studio audio engineering. When producing sound for "the house" audience, one uses:

> a sound board . . . power amplifiers that amplify the speakers, so there's a left side and a right side . . . many microphones & accessories . . . direct boxes – [so that] if a guitar player has a pickup on their guitar and they just plug right into this box, and it goes to the sound board without the use of a microphone . . . graphic equalizers – to play with the tone of the sound . . . effects [such as reverb[eration] used on a vocal or guitar . . . vocal compressors . . . to keep singers from getting so loud that there is distortion. A monitor engineer uses onstage [in the back, behind the curtain] a sound board made for monitors, power amplifiers which power the onstage monitors which are in front of the musicians so they can hear themselves. There are different mixes for each musician on stage based on their individual needs, and the same monitor engineer does all of that mixing; each mix has a graphic equalizer attached to it to help reduce feedback, and that's a skill.

She further conveyed how studio technology has changed:

> There's a mixing board that's designed more for recording than for live sound . . . same concept, but designed a little differently . . . we used to have analog tape machines . . . with 2" wide tape and 24-tracks available, and [on] each one of those [reels of tape] . . . you only got 16 minutes and it cost $175 a roll, so those were the days when we spent a lot of money on tapes, but those days are gone … you'd need 3–4 tapes to do a whole album . . . running 30" a second with 2400 feet on a reel. . . . Today you have . . . usually a Mac computer – most studios run Macintosh, and we run external hard drives and Pro Tools recording software [which is] the industry standard [though] Logic is [another] that is . . . popular in the mainstream industry. The rest are very consumer based – like Sonar, Asset Pro, Digital Performer – a lot of those are . . . used in home studios. And the rest of it in a studio is similar to live sound: . . . microphones, power amplifiers to amplify the studio speakers, only sometimes we have two or three sets of speakers so we need enough power amps to power more than one set of speakers, compressors, all kinds of effects. But the recording software gives you a lot of those tools within the program, and you can buy more. They're called 'plug ins' and you can add a compressor to a track, or add a reverb. [In the studio, the musicians use headphones instead of monitors] to hear themselves and the already existing recorded tracks; [you] can't use monitors in a studio because the microphones used to amplify voice or instruments would pick up the sound coming out of the monitors.

Skills involved

Clearly, audio engineering – whether in live performance or recording studio settings – is highly skilled and varied, making it all the more attractive to competent, creative women. Boden revealed that she likes live audio engineering because "it's very eclectic," adding:

> You have to bring in many different skills. You have to understand the acoustics of space, you have to understand how the equipment works, you have to know how to mix, so therefore you have to have a musical ear, and then you have to do well with people because you typically . . . have a sound check and interface with the artist and have to get their needs met for their kind of sound . . . you have to work . . . the people who travel with them. And so it all has to come together in a very short period of time, 'cause typically live concert sound . . . is not a permanent installation – it's temporary.

Task variety is evident in studio work, too. Leslie described her additional administrative responsibilities as Director of Music Recording and Scoring at a major audio studio:

> running the recording studio – which involves budgets, employees, booking time, marketing the studio, and record production in which she works with the artist in the making of their record [including] song selection, picking musicians, arrangements, and engineering the sound or working with another engineer.

If a record label is involved, she also coordinates between the artist and that company.

Skill can be thought of in part as the knowledge and judgment needed to solve problems in the face of unforeseen challenges (Haddad 1989), and the work of audio engineers certainly exemplifies that definition. Christine offered examples of the types of challenges that are encountered in live sound production that go beyond losing the signal from an instrument: "Lots of different instruments, lots of open mikes . . . in a festival setting it's having 6 full bands back-to-back, that you have to change for each one." Weather is an unpredictable factor when working outside, and "we've had mice build nests in the back of our console and chew the insulation on the wires."

Motivation to enter the field of audio engineering

Although the career paths followed by the women differed to some extent, there are common elements. All discussed a love of music as a key factor motivating their entry into the audio engineering profession, though in each case their discovery of that occupation was circumstantial. Three of the women reported having had some formal musical training in their youth and came from families whose parents (one or both) played musical instruments. While those experiences gave the women childhood exposure to music, it was not necessarily in a mentoring way.

Christine's father had been a "big band" leader in his youth, but realized that was "a tough way to make a living and became an oral surgeon instead," playing classical string bass and tuba on occasion with a local symphony orchestra. Her mother had been a classical pianist who had her own radio show prior to marriage. Later as full-time homemaker she "forced" her daughters to study piano "at a young age." Christine ultimately chose the flute as her formal instrument, minored in flute performance as an undergraduate student, and served in her college's marching band and symphony orchestra. She also played keyboards and electric bass which she "picked up" on her own because she had friends who needed a bass player in their band. Since she had "played a little guitar" she thought she could learn the bass and "just started playing" it.

Her undergraduate major was psychology. While she had excelled in high school math and loved science, she encountered gender-based obstacles that limited her pursuit of those fields:

> it wasn't really fostered; it was sort of looked on [by her teachers] as a curiosity . . . and . . . when I got to college . . . I was actively discouraged from a science major, which now . . . makes me unhappy, because it's what I probably should have been doing in the first place.

Neither was her strong mathematical ability encouraged at her high school. Christine recalled that a female math teacher once admonished her for answering questions, stating: "you make the other kids in class nervous." She chose not to major in music because "I wasn't sure I could make a living at it." Yet because she attended George Peabody College in Nashville – a music-rich area – she was able to obtain work in recording studios as a musician, playing flute for commercials, corporate jingles, and also serving as a recorded background musician for industrial films.

That exposure to recording studios, combined with her love of science and math, triggered a curiosity about recording:

> I just got interested in what was happening on the other side of the wall from where I was, so I started hanging around and just watching the people who were actually doing the recordings instead of being on just the other side of the microphone. And . . . I was fascinated by it.

She was able, through a friend, to make contact with the head engineer of live broadcast at the "Grand Ole Opry," so she "got to hang around that stage while they were putting that on . . . [and] to experience a little bit about what happens – what it takes to actually make that [live broadcast]."

Audio engineering also fit Christine's personality and intellectual interests. She characterized herself as "really shy" which made performance difficult for her. She enjoyed playing bass because she believed that "nobody pays any attention to the bass player." Operating a sound board allowed her to be "an integral part of making music happen, and I love being a part of making music happen, so it was a way to have an impact and to be part of something I loved to be part of." Given her longstanding interest in math and science, the technical aspect of audio engineering appealed to her as well: "I wanted to understand what was happening." Audio engineering for her is "a way of combining art and technical knowledge."

Christine would go on to earn a doctoral degree in school psychology and while in graduate school became the first female manager-level person hired at the university's computing services center, in charge of the user services group in academic computing. Ultimately she became a professor of educational psychology at a Midwest public university, continuing to work part-time as an audio engineer at a highly regarded local folk and acoustic music club. Some years later she would elect to give up that tenured faculty job to pursue another graduate degree – a Ph.D. in neuroscience, which she has since completed, even while continuing to do part-time audio engineering at the aforementioned local music venue and at women's music festivals.

Liz's parents also encouraged her to take piano lessons during her childhood. Her mother played the piano and taught at school (until her marriage, and resumed her career later), and her father, who earned his living in the skilled trades, played saxophone in a "big band" and French horn. Her five siblings also studied piano, and one brother studied cello as well. She had "done some theatre in junior high and high school, and . . . had some familiarity with the stage, and I really enjoyed it". She reported that the "familiarity with the stage from younger days, and interest in performance and music" is what "drew" her ultimately into the field of audio engineering, though the path for her was not a direct one.

Her athleticism and enjoyment of high school sports (field hockey and basketball), perhaps encouraged by a field hockey coach that she regarded as "a strong female role model," had led her to study physical education in college. Yet she found college life dissatisfying and dropped out after two years. When she did resume university studies at the University of Massachusetts (U-Mass) Boston a year later, she chose psychology as her major and selected classes with "radical" professors like Linda Gordon and David Hunt in fields that appealed to her intellectually and politically:

> I was picking the good teachers . . . and taking revolutions in modern history, and Russian literature, and women's studies, although at that point we had to call it 'the history of feminist thought' – it wasn't even allowed to be called theory. So I was meeting radical feminist women [in 1974–76] and it was thrilling.

She pursued her studies full-time, doing a bit of freelance audio work on the side, and graduated *magna cum laude*.

The freelance work came to her indirectly through a girlfriend who had obtained a job at an audio production company thanks to a U-Mass Boston program called "Women in Career Options." Liz "went with her to a gig for her company" and discovered that the performing feminist duo – Jade and Sarsaparilla – was looking for a person to mix their sound. Since her girlfriend was not available, Liz volunteered her services and the duo taught her what they knew.

She continued the freelance work after moving back to Maine for personal reasons, while also taking a laborer's job at a paper mill. Three years later and back in Boston, Liz landed a job at the audio company where her girlfriend had worked as a result of the Women in Career Options program – Terry Hanley Audio Systems in Cambridge. One of her first assignments was to drive a truck to New Orleans and work on the stage sound for the New Orleans Jazz and Heritage Festival in 1979. Liz described that experience as:

> . . . mind blowing, because it was black and white people doing music together. And it was amazing music – all those wonderful New Orleans artists. That was just super exciting and fun, and really made me want to continue. I wanted more involvement.

After working for that company for five years, she began her own, with a client base built up from her freelance days: "And 'cause the women's movement was happening and the Viet Nam war . . . they wanted a progressive and/or a woman to do sound even if you didn't know anything. That was basically how it [her career] progressed." She didn't buy equipment until she went into business herself, and she "started small and ended up having to buy a lot."

Boden, too, had studied music in her youth – the French horn – and like Christine loved and excelled in science and math. She reported that her parents wanted her "to do well" in school, but weren't particularly encouraging of her interest in science and math, since:

> like most parents back then, their goal was really for me to go to college and get married. That motivation [for science and math] came from me and my teachers . . . I was just the type of kid that had to succeed in everything . . . I was kind of driven to be the best in everything.

Ultimately Boden's interest in science, like Christine's, would be thwarted by gender-based obstacles placed in her path. Having graduated high school as class Salutatorian and the top female student in her class, she earned a scholarship to study at St. Lawrence University. She sought to major in physics, and was placed in advanced (second-year) physics and calculus courses.

> But I was in over my head, and that had a lot to do with sexism in a way . . . I might have been able to cut it if I had been a boy, because the girls had curfew, and I had to be in my dorm at 10:00 at night, and all the guys would get together at night for my physics class and do their labs together. And I'd have to be . . . on the payphone in the basement trying to get some guy to talk to me about what they'd learned, and it was very hard. And

none of the guys wanted to be lab partners; I was the only girl in the advanced physics class, so I finally succumbed; I dropped that [physics] as a major and I ended up majoring in English 'cause I like to read . . . and I minored in music. I would have majored in music probably if they had it, but they only had a minor.

Boden would ultimately earn a graduate degree in library science from the University of Michigan, move to Boston where she did library and political activist work, and relocate to Washington, D.C., working at the Martin Luther King library. During that period she realized that:

I really missed music. I had played French horn all through college, and I stopped playing once I got out of grad school. . . . I really literally sat in the library and thought about how [would] I get music back into my life. I missed it terribly.

Yet her work schedule and political activism left her little time to practice the French horn, leading her to realize that she could not play it professionally; moreover, at that point she did not even own one. Then something fortuitous occurred:

I went to my first women's music concert [at George Washington University] , and saw Judy Dlugacz [a member of the Olivia Music collective] mixing onstage, and I really wanted to learn how to do that. . . . it just clicked – 'that's it; that's what I want to do' [it] looked like so much fun, and I could just tell that she [Judy] had something to do with how it sounded. I had no idea [that would become] my creative outlet. . . . And she [Judy] was the one who . . . told me if I wanted to learn how to mix to seek out Casse Culver because she knew Casse was looking for someone to train, 'cause she was tired of trying to sing . . . and run her own sound at the same time.

It took Boden a year to link up with Casse, but once she did, "that confirmed the fact that I really loved mixing. To me, mixing was my art. . . . I think of mixing sound as being like a conductor, and so it really satisfied that need I had to be a musician. . . . "

The District of Columbia was fertile ground for the launching of her audio engineering career, for it was rich with women's music culture in the mid-1970s.

[I] started mixing with her [Casse's] little PA . . . at a local nightclub – Club Madame, and that's where I got to know all the different women's music performers, as they came through to perform there, and because of that I started to get hired at women's music concerts and local political concerts . . . [and] in different churches and the women's center . . . particularly

All Souls Church . . . that's where I met Sweet Honey in the Rock. And then more and more people started to hire me, so eventually Casse and I decided that what was needed was a women's sound company, and there were some other women there who were doing sound . . . [such as] Lee Garlington. . . . Casse and I decided to start a business . . . a company [called] Woman Sound, and we rented our equipment in the beginning.

However, rental equipment proved unreliable, so they eventually bought their own:

Then Woman Sound . . . rapidly got a really good reputation. Because DC's such a political town we got hired by many different political groups, and for political functions. But that led us into other areas of DC, like the Smithsonian . . . and we literally eventually became the sound company for DC in many different ways – we did all the mayor's gigs, we did almost all the major festivals . . . [and other] interesting musical productions. . . .

Ultimately, Casse's involvement with the sound company lessened since singing was her priority, and Boden bought her out. She became "known for quality mixing" which she attributed to the fact that as a woman she was "sensitive to people's needs and [would] listen to musicians" and because of her "classically-trained ears playing in orchestras and bands and . . . quintets for years – I really could hear everything, and just had the knack of being able to make it be so in the room, so that the balance was something that everybody enjoyed, which is really important."

Even those interviewees who did not report extensive formal, classical music training expressed their love of music, and learned to play guitar well enough to perform with local groups. Karen, who grew up in Queens, New York, stated that she was interested in studying a musical instrument as a child and began taking flute lessons in eighth grade. She described herself as an "ok player" and enjoyed it a great deal, stating: "I didn't get a lot of encouragement from my family in the arts, and so it really . . . was a self propelling thing. If I wanted it, I had to go after it." She added: "coming up through the hippie era, I really didn't like school." Because she wanted to "hang out in Greenwich Village and play music," she chose to earn a business diploma focused on typing, steno-graphy, and related skills rather than pursuing the college preparatory track in high school.

After graduating at age 19 and attending the Woodstock Peace and Music Festival in mid-August 1969, she moved out of the family home and spent a good deal of time in Greenwich Village. Although she loved to play folk and rock-and-roll rhythm guitar, she never felt that music performance would be her career. She did recall, though, a childhood interest in music technology. When she was 8 or 9 years old the family got a small tape recorder "and I was fascinated by it; I took it apart, put it back together . . . so from an early age, there I was playing with tape recorders." Throughout her childhood, she did not encounter female role models possessing technological proficiency: "I cannot recall seeing any women doing anything but have children."

In speaking of her parents, Karen stated that "the only thing they saw for me was a husband and children, which of course, I never gave them." Still, she had the freedom to pursue her own career aspirations, and was not pushed by them in any particular direction. Her father saw her as smart and able, and knew that she loved music. He learned of an opportunity for Karen to obtain a job as an assistant business manager at a recording studio that produced commercials (advertising jingles) working for the "head honcho studio manager," the daughter of a family friend. Although that woman had initial reservations about hiring a friend of the family, she enjoyed the interview with Karen and deemed her highly qualified for the job. "It was actually my dad who was responsible for getting me into the business. . . . When I got that job and started being around recording equipment, I just fell in love with it."

Still, there was an "unspoken rule: 'women didn't touch the equipment' . . . it was a given that women were in the business end of things in studios and women weren't supposed to be engineers." Karen became studio manager after a year, and ran the day-to-day business operations. After a while, though, she ignored the unspoken rule and "started playing around with the equipment." Having worked for three years at that studio, she wanted to learn more about music, and moved to Boston to attend the Berklee College of Music. She found the atmosphere of that school "very sexist" and not oriented to the musical styles of guitar that she wished to pursue, and left within a year. Ultimately, she discovered that "the [recording] studio became my instrument."

Like Karen, Leslie was a self-taught guitar player. She was the lead and rhythm electric guitarist in an all-women band in which her partner served as the drummer. At a younger age, she had performed in an acapella vocal group with two female cousins and a male friend, singing music composed by a male cousin. They were well known in the Los Angeles area, and "spent a lot of time recording masters in recording studios," but instability of labels in the record industry was an obstacle to production of a record for the group. Her father had been a musician and band leader and her mother a vocalist, but her family unit was disrupted by the early death of her father and her mother's out-of-state move following re-marriage three years later. Leslie chose to remain in Beverly Hills to finish high school, and lived with her aunt, uncle, and the cousins with whom she sang and spent a great deal of time.

Her first exposure to recording work followed the dissolution of the all-women band in which she had performed. After her drummer partner joined another band, Leslie:

> started arranging their backgrounds and mixing their sound. And that . . .
> led me to . . . understand that . . . I had the power – power is such a funny
> word to use – that I had the ability with sound to affect the way I heard

things. So if I thought that the backgrounds were out of balance, I could fix that by balancing it the right way at the console. And I found that I had an immediate aptitude for that – kind of just like I did playing guitar, but I was self taught and played guitar for many years, and here I was at the beginning of sound stuff. And I thought that . . . I could actually try to learn what it is I'm doing. So I started a PA company with a couple of male friends of mine, and we pooled our equipment together, and I was mixing sound, and doing that, and I had a small studio in my home because at that time it was hard for people to afford recording equipment. If you wanted to do a demo, you had to go into a studio, and then I took a couple of classes just to . . . have somebody to ask questions of also.

As was the case with all of the other women audio engineers interviewed, Leslie found herself drawn to audio engineering because "I saw it as an extension of my musical and creative ability and desire."

The common elements discussed thus far that attracted each of the women into audio engineering were: 1) a love of music; and 2) technological aptitude and interest. An additional shared factor was that most of the women strongly identified as feminists, and saw their mastery of audio engineering technology not only as a way of earning a living (all were self-supporting), but also as a vehicle for social change. Audio engineering was an exclusively male domain – one not easy for women to enter since informal apprenticeship was the primary mechanism for knowledge transmittal. Yet given the activism of the era, there was broader support from networks of women regionally and nationally – not only among the few women audio engineers, but from the broader women's music and feminist communities.

Learning the craft in a male work culture

On-the-job learning

With the paucity of female audio engineers in the industry, women who were willing to work their way up and prove themselves through long evening hours and heavy lifting (literally and figuratively) were not necessarily regarded as threats by their male colleagues. Since none of the women I interviewed sought to enter a male industry with dating in mind, boundaries were clear and women were more likely to be respected for their abilities than for their physical appearances. As Karen put it, she knows "how to be one of the boys; I dress [in] jeans and t-shirts like they do, and I . . . have a way about me that is not distracting as a female." She gave an example of the opposite of that – once while working in Toronto she hired a female assistant engineer to help her with an album. The woman was "gorgeous" and "decked herself out when she came to the studio . . . and the guys couldn't take their eyes off of her. It was too distracting. So I had to ask her to dress down for the next session."

It didn't hurt, either, that women were in some cases required to work at lower pay levels. Liz noted that when her girlfriend was hired into the sound company through the "Women in Career Options" program, as described earlier in this text, the owner "had an openness because I think he paid her half of minimum wage, and the university paid her the other half." When Liz was herself later hired into the same company it was at "minimum wage." There was no formal apprenticeship or in-depth training;

> it was all seat-of-the-pants. . . . The guys showed me some things, definitely, but it wasn't really a teaching culture – even among the men. It was a watch-and-learn culture. There was some instruction, but it was mainly watching and doing; figuring it out.

Once in the studio, women worked their way up. In Karen's case, after leaving Berklee College of Music, her past studio management experience enabled her to secure a position at Intermedia Sound studio in Boston. Although she was once again hired as a manager, she continued to be drawn to audio engineering: "I started recording right away; I started playing with the equipment nights, weekends, and I started teaching myself. I had access to the studio, and I was allowed to use the studio whenever I wanted."

Eventually she approached the owner of the studio and asked to be "fired" as a manager and rehired as an apprentice engineer, and he agreed to let her switch jobs. She took a pay cut to do so. She began by being an assistant to the engineers – moving microphones, and even cleaning toilets, and she worked her way up to being a full-time engineer at that studio, "thanks to him."

> By the time I became an apprentice, observation was key, asking questions of the men who were my mentors at the time. I spent a lot of extra time . . . late at night at the studio, hanging out with the guy who used to fix all the things in the studio. I'd hang out with him all night long, asking questions and playing around. . . . My whole life was that studio. I was there during the day, and I was there 'til late at night. So being around it – it's osmosis and asking questions. And there was only one book called Modern Recording Techniques written by Robert Runstein and that book was my bible. . . . . I had the opportunity to ask questions and read, and then go practice it. . . . Essentially in the early '70s I was self taught.

Mentors and formal courses

Although Karen described her early learning as being self-taught, she also had male mentors at the studio who were "really good engineers and very knowledgeable" with whom she was able to "sit and talk." And if she didn't understand something by either trying or reading it, they would give her more information. She used the term "mentors" to describe them "because they were also supportive."

In 1976, the New-York-based Recording Institute of America began to offer audio engineering classes on an experimental basis at the studio at which Karen was employed. They hired the engineers there to teach courses, which Karen got to attend for free. She would practice what she had learned in class in the studio at night, and completed the introductory and advanced courses. She was an anomaly, as "the only woman in classes all the time."

In 1977, after she had been a full-time audio engineer for a year, Intermedia Sound studio was sold, and:

> the new owner came in and laid everybody off who was involved in the studio, and brought in his own people. So I went to every studio in the state of Massachusetts for a job as an engineer. And I got two offers – both to be a studio manager.

She declined them, and took out an ad in a local publication called *Musician's Magazine* offering her services as a freelance engineer. She ran the ad every month "for years" resulting in "more and more and more work" including live sound as well as recorded albums. She became one of the early successful freelance engineers in Boston of either gender, and from 1977–1990 "really honed" her skill and "did a ton of albums." By 1981 she limited her live sound work to focus more exclusively on studio work.

Karen relocated to Toronto in 1990 "for love" and began to build up a client base for freelance studio work once again; during the 12 years she spent there she attended courses at the Harris Institute for four years to learn how to teach audio engineering to others, and found that she "loved" doing so. Moving once again in 2002 to her current location of Wilmington, NC, she added "audio educator" to her skills of recording and live audio engineering, and now serves as a mentor to other women (and to men), teaching courses on various aspects of audio engineering, including theory and application, live recording, mixing, and in-depth work with the Pro Tools® digital audio workstation platform. She reported having had her "best year in business ever" in 2008, the year before I interviewed her.

Leslie's description of her career progression in audio engineering also reflects the watching and learning culture that Liz described, and the mentoring opportunities that Karen experienced:

> When I started [in 1975] our industry was quite a bit different than it is now. Most people that got jobs started out as apprentices. Most of the studios were owned by record companies. Many of the studios were union. And there were very few independent studios around. So when I started I got a job at ABC studios [in LA] as a production engineer, my job was really to make tape copies. I did that for about 6 months, until I volunteered to be an assistant engineer on a project. Then I was in the studio working with another engineer on their project. And I had already had some experience on my own; I owned a PA company and I had mixed live

sound . . . , but certainly not the kind of experience one would get by working in a studio. So when you work in a situation like that you are next to whoever the first engineer is and you're in some ways mentored by them, whether officially or not. You spend a lot of time learning by watching how other people do what they do, asking questions, and I was always mentored by people all along the way. . . . I was in a position to assist really great engineers, and I was mentored by them, and then I started doing all my own projects.

When I asked Leslie whether she had encountered gender-based obstacles in her work, she replied in the negative, though she did acknowledge that when she initially sought work as a recording engineer as a step toward her ultimate goal of becoming a record producer and manager, the studio director's response was that "he didn't know how the clients would react to having a woman on the recording sessions, but that he would hire me and we would see what happened." She also recalled an incident in which she was "actually removed from" a session "because the guy's wife found out I was in the control room, and she didn't want any other women in the control room". Still, she felt that throughout her long career in the industry:

the only obstacle was my own lack of knowledge, and if I took the opportunity to ask the right questions and keep learning, then I would be providing the kind of service to my clients and the people that I was supposed to be working with as good as I wanted to. When you're the assistant engineer on a project, you're really supposed to not only be the interface between the recording studio and the client, but you're supposed to know everything about all the equipment in that room. So in order to do that . . . I would always be raising my hand saying I had to go to the rest room and instead I would go to the tech shop and ask the guys questions, and then come back. The poor engineers I was working with probably thought I had quite a severe bladder problem.

Christine also began to learn audio engineering by apprenticing herself to people and "following them around." Following an academic year in Israel, she stayed on for an additional two years (1979–1981) to work with a psychologist with whom she had studied in the U.S. While there, she became a member of a band that played all over the country, and performed on the radio. She also became an assistant to an audio engineer who did live recordings of music performances for broadcast on an Israeli public radio station. She followed him around to watch and learn, plugged in cables when he asked her to, and she "asked a lot of questions. He was very patient explaining things to me."

Although male audio engineers were willing to mentor these women, clients did not necessarily feel the same way. Karen recalled that:

> One time . . . in 1978 . . . I was at a studio [freelancing], sitting behind the
> board doing engineering . . . , and it was almost time for the next session to
> start, and the client for the next session opens the control room door, looks
> me straight in the eye . . . , and says 'where is the engineer?'

That simply made her more determined to pursue a career in audio engineering.

Additionally, some male engineers were not as accommodating as those with
whom Leslie, Karen, and Christine had worked. Although Boden had entered
the field of live audio engineering through her exposure to producers and performers of women's music and her work as a disc jockey work for a feminist
radio program – *Sophie's Parlor*, it wasn't long after forming a women's sound
company that she began to realize the limitations of her knowledge. As they
began to obtain political "gigs requiring bigger systems," seeking help was
"very difficult."

> I called so many different sound companies in the area and it was totally a
> male-dominated field . . . all guys, and I wanted to apprentice with them or
> work with them and learn, and none of them would take me seriously and
> none would let me gig with them . . . because I was a female. But there was
> this one company – National Sound in VA that did say 'come on out' and
> they were really big; they toured with Rush and there were 2 guys who
> owned it [Greg Lukens and Tommy Linthicum] and they were just fabu-
> lous, they were my mentors throughout and became best friends. But I did
> have to load a semi truck with huge speaker cabinets into the middle of the
> night to kind of prove myself, but once I got past that hazing they really
> wanted to help teach me and Casse . . . . They didn't really teach us much
> there, it's just that they took us seriously and would rent us equipment.
> And they would tell us how to hook it together . . . but it was very much
> on-the-job training.

She recalled doing sound at an early Meg Christian concert at All Souls Uni-
tarian Church (in Washington, D.C.), and things "didn't always work the way
that the guys had told" her they would. Since cellular phones were not yet in
use, she would have to dash down to another area of the building where there
was a pay phone and call Tommy or Greg for advice on what to try. This
would sometimes have to be repeated if the advice they had given did not solve
the problem:

> That was typical in the beginning, because it was just totally on-the-job
> training and because the guys always had such an advantage 'cause they'd
> be hired and they'd go out all together and learn from each other. Whereas
> we just got set up with this gear and had to go out and do it. . . . I just
> loved it, though, so I would just remember what I finally did that made it
> work, and then you'd do that the next time you were out. It was very much
> trial and error, and reading. And I took a couple of courses – whenever

some kind of workshop would come along that was appropriate I would take it, and that helped.

One such workshop was a week-long program offered by a company called Synergetic Audio Concepts, and developed by husband–wife team – Don and Carolyn Davis – which covered electronics, electricity, acoustics, and was designed specifically for audio engineers doing installations. She had learned about it from the fellows at National Sound who knew that the course was coming to D.C. and they recommended that she take it. "It was really intense – very difficult, but very worthwhile." In fact, Boden stated that she used the textbook from that workshop throughout her career.

  She in turn used her knowledge to mentor other women: there were "so many women who wanted to work with me. . . . I started a course on how to do concert audio engineering" for women only, initially, which she held in her apartment and at All Souls Church, and used the aforementioned book as the basis for the course.

> And then eventually in the middle of my women's sound career, I got a Master's degree [MS] in audio technology at American University [1986]. What I really didn't have was a knowledge of electronics and electricity, and I was at the point where I really needed to know more to really understand how it all hooked together and how it worked.

American University offered her an opportunity to teach her course and run their electronics lab in return for getting her tuition paid, which she said "was great" because tuition there was "very expensive." She enjoyed the program of study considerably because it was "very eclectic – I took lighting design, and video . . . , and electronics and electricity, and acoustics, and then I could take a recording class . . . and that was excellent."

  Most of the women took formal courses or attended workshops to supplement what they learned through informal apprenticeships, out of a desire to learn and understand theory as well as practice. In addition to Boden and Karen's aforementioned studies, Liz took electronics courses at MIT's Lowell Institute, and at the Northeast Institute on Beacon Hill, and Christine attended a program at The Recording Workshop in Chillicothe, Ohio for four–six months. Leslie did a good deal of study on her own. This is indicative of the seriousness with which all of the women approached their work, and their recognition of the need to be the best that they could in an industry in which they were so few in number.

Confidence levels

Despite the obstacles that some of these women had to overcome to learn the technical and applied aspects of audio engineering, all of them reported fairly high confidence levels. Two measures were used in the interviews:

"Were you confident that you could learn audio engineering? How confident on a scale of 1–5, with 5 as the highest level?" The reported scores for each of the five women were: 3, 4 (at first, and then 4.5), 4.5, 5, and 5 again, with 5 representing "extremely," 4 "very," and 3 "somewhat." In explaining her 3 rating and the evolution of her self-confidence leading her to being a very highly regarded audio engineer in the industry, Leslie stated that she was fairly self-confident as a person "in the way one moves through the world,"

> but I was not very self confident about what I knew. So that's why I would characterize it as a '3'. . . . I always had an innate ability around it, but there were certainly challenges for me from a technical perspective, so I had to work very hard at that. I had to read a lot, and I still do, to this day. . . . The industry I'm in both from a recording standpoint or just from a business standpoint changed constantly. So I spent a lot of time trying to learn more about what it was I was doing by talking to people, by reading magazines, by sometimes just nodding when I didn't understand a word they were saying; trying to spend a lot of time and energy exposing myself to as much as I could. To do what I do and what I was doing then, there had to be a certain amount of self confidence – being very focused, and trying to understand what you were doing and if you didn't, trying to figure out how to get the information that you needed to do the job that you were trying to do.

She also recognized that as a woman in a predominantly male industry, she needed to prove herself by excelling at her work:

> Part of [my motivation] I recognized early on [was] there were already assumptions about me and my abilities by other people – based on gender, so trying to have a certain level of self confidence about what I was doing kind of made that go away. . . . So that kept me learning more, reading more, wanting to absorb more, spending as much time as I could trying to figure out what I was doing – also because I wanted to do a good job as an assistant engineer and wanted to learn more. But in order to master that, you spend 8 hours a day in a small room with a bunch of guys. You have to in your core, I think, have a certain amount of self confidence as a person.

Karen, too, saw her confidence rise over time. Initially she would have rated herself with number 4 – "very confident" – because when she was working at the studio, she hadn't realized how difficult it would be to master audio engineering. After recognizing that "this is here for me to learn, and I'm going to learn it" and pursuing that knowledge, her confidence level "got better and better" and rose to a 5. She conveyed what accounted for her belief that she could learn audio engineering:

The confidence came from a passion for the work; a passion for music; a passion to get myself [into] something I could make a living out of, because I knew that [in] playing music I wasn't going to be any great musician. And so out of passion for music and from being around all that recording equipment and thinking '. . . I like this; I think I can do this; wouldn't it be great to have a career doing this.' I think passion is the answer.

Christine, who rated her confidence level at 4.5, described herself as "quite confident," adding:

there isn't anything magical about it. I think the art is harder than actually learning how to deal with the equipment itself – getting the mix to have it sound right so that everybody's happy – the artists are happy, the audience is happy, everybody is happy. Actually physically being able to use the console is not so difficult.

She added: "I started out as a musician, and I think that's helped being an engineer as well."

Liz rated her confidence at 5 not necessarily because she had great self-confidence throughout her life, noting that "all girls grow up with a certain amount of lack of self worth . . . even with some lack of self confidence one can still persevere and push through because there's . . . a higher agenda or goal." She added:

I was a good learner and I watched what they were doing, and I was later very much influenced by musician friends who really taught me how to listen musically and not just technically. And that was a huge influence on me and quite a boon to my career because I feel to this day, the technology really just needs to facilitate the musical performance. It's bringing culture to us. It's an art form in and of itself. I think good engineers are technical artists for sure, [but] it's different than just turning things up and down. There's really a high degree of difficulty and there's a skill level involved, and a bit of good luck and grace needed.

Boden also rated her initial confidence level as "5", due not only to her past academic success, but also because of the strength of the women's movement during the 1970s and her connection to such:

it was the period – because of feminism – women, including myself, were feeling that they could do anything. There were so many models for that around, and particularly in DC because of the Furies Collective [a lesbian separatist organization] . . . the whole political environment in DC was for women to set up their own businesses and take control of their lives, so everyone was making things happen

And so she did by establishing Woman Sound with Casse in 1975, Two other illustrative examples she mentioned were Lammas Bookstore and Roadwork. The former was a women's/lesbian bookstore and organizing venue that operated from 1974–2000 (Hoover 2000). Roadwork was founded in 1978 by Bernice Johnson Reagon and Amy Horowitz "to put women's culture on the road" in the form of music festivals and concerts featuring political artists and organized by activists (Roadwork Center 2017, 1).

Individual and societal benefits of their work

Liz stated in response to my question about whether or not she viewed her work as part of a movement for social change (the women's music movement), and whether that was a primary or secondary motivation:

> Definitely, because I was a young activist, and I was involved in the anti-war movement, and I was involved in the women's movement. It's hard to differentiate whether it was a primary or secondary motivation. I felt like my life was about social justice at that point; so it's all part of the same package.

Boden's prior political activism had given her a sense of efficacy as she worked in socialist, abortion rights, and peace organizations, but it was feminism and the women's movement that changed her life, and led her to discover women's music:

> one reason I left organized politics was because there was a 'glass ceiling.' And I was more interested in the women's movement . . . and was involved in the women's rights movement, and I got attracted to women's music. [Women's music and social change activism] went hand in-hand because I wanted to have music back in my life, but really where the confidence came that I could do [audio engineering] was from the women's rights movement, and feeling that I could change my reality as a female doing something like that. I didn't need to stay and be a librarian in a female-type job that I liked well enough but wasn't passionate about. And that I really could do more with my life, and that's part of why I left the party, too . . . I learned about political rights and quality of life, and I wanted to act on it as a female for myself.

Moreover, the expertise she would gain in audio engineering became not only a career, but a way of giving back to the women's movement:

> And it was a conclusion at the time . . . that 'the personal is political' and one had to live your politics. Women in the left were the givers . . . [and thought] 'is it ok if I stop giving, and taking for myself and doing for myself?' and the 'personal is political' [suggested] 'yes' because then that would benefit not only myself, but would benefit the world and women

greatly because I would be a whole person. And that was really a political philosophy that many people were putting forth in the women's rights movement . . . and it turned out to be true, because not only did I move myself forward in life . . . but I helped women and political events here. . . . and that was a really big motivation for Woman Sound – it was . . . one of our mottos – to . . . allow everyone who was working for a cause to have the right to have whatever they were saying heard, and they weren't before that.

That the audio engineers I interviewed were willing to give back was evident not only in their enabling of women's music to be heard, but also in their mentoring of other female audio engineers. Examples are the aforementioned courses taught by Karen, and Leslie's recording instruction at the Institute for the Musical Arts – a teaching, performing, and recording facility supporting women and girls in music established by legendary guitarist June Millington and her life partner Ann Hackler.

Another is women working together on live audio engineering at festivals and events. Christine spoke highly of her past work with a more established feminist live audio engineer, who had "high standards, there was help [provided] in getting to the standard. I appreciated it; I learned more about professionalism from that than anything, and I still . . . have that as my habit of working."

Christine further indicated how she derived benefit from the women's music movement even though she entered it already trained as an audio engineer:

[women's music] gave me opportunities that I don't think I would have had otherwise. . . . I did identify with women's music, and I did think it was something special that we were creating. To look at a place like Michigan where all the carpenters, all the technicians, all the engineers, everybody – they're only women – and to make a community like that happen with all of its various parts was extremely empowering. And for me as a woman in a highly male-dominated field – that's where I got my chances to work, because I don't do it as a profession . . . and even if I did do it as a profession, it would have been hard for me to get the kinds of opportunities in the larger society that I was able to get in the women's music community.

For me it was also the level of professionalism; it wasn't that we were just sort of putting something together and we were excused because we were women, or we were grassroots . . . that's not an excuse to be sloppy. . . . There was a level of expertise that was expected and nurtured in that environment as opposed to excuses being made. I think a lot of times in the folk community you hear, 'well, it's good enough for folk.' And particularly . . . the group of us who were involved in the technical side of it took a huge amount of pride in being just exquisitely skilled and good at what we were doing.

Christine's skill was recognized a few years prior to this interview when she was given the technical contributions award from the National Women's Music Festival.

Leslie, too, entered women's music work after she was well into her audio engineering career, and described the mutual benefits derived from that association:

> I was coming to it [women's music] from a very professional perspective where I had already been working as an engineer for 6 or 7 years, working on all different kinds of projects, as opposed to some of the women's music projects where they brought in people to do the engineering that were not necessarily working in professional studios on regular recording projects. So it was really quite a wonderful experience I think for both sides because I could bring my expertise and record making and professionalism to their projects, and yet for me it was just fun to hang out with women all day. June Millington and I had already been friends, and she produced the first record with Cris Williamson that I did.

Although Karen's work also goes much beyond the production of women's music, she, like the other audio engineers, recognizes and appreciates the extent to which that movement has "absolutely" contributed to social change:

> that's what the festivals were about. They . . . were started because especially lesbians but all women wanted a space where they could be without men. And the festivals are not as well attended today because it worked – there's women's spaces all over the place. And the pioneers early on made it happen, so it's not as needed as it once was. . . . Doing sound at a music festival is stressful, but it's lots of fun. So the motivation was doing work I love, getting paid, and having a ton of fun. And the social change aspect of it – that was already there. . . . engrained in the work by virtue of the festival.

Each of the women described the benefits they have derived from their audio engineering work. Liz stated: "Primarily I've heard a lot of wonderful music, and had the opportunity to work with tremendous musicians and dancers and some actors." Karen remarked that:

> The biggest benefit has to do with my whole life . . . since I've been 19 years old when I started out at 6 West Recording in New York City . . . I've been able to do work that I'm passionate about. So many people in this world have jobs they hate. I've been able to make a living doing work I love and that I'm passionate about . . . What bigger benefit is there? [Another benefit has been] all the wonderful memories and musicians that I get to spend a lot of time with . . . [some of whom] I became close to and

have maintained friendships with. . . . A large percentage of my clients become friends . . . Wonderful!

Leslie described the benefits she has gleaned from audio engineering in this way:

> It's really given me the opportunity to express my own creativity in a way other than being a musician or singer, which is how I started when I was a teenager and eventually got into sound and stopped playing. . . . It really gave me the opportunity to express my creativity in a way that I kind of couldn't do as a musician because I was self taught. In order to be really good at that I would have had to have gone back to school and spent a lot of time studying and playing. And that wasn't the career that I sought. And there is, I think, something really great about making a contribution to a record and supporting somebody's art, and being a recording engineer or a record producer really enables you to do so.

Boden spoke glowingly about what she had gained from her work as an audio engineer.

> I had the benefit of becoming creative all those years, I had the benefit of becoming very well known, I had the benefit of being totally immersed in Washington, DC events . . . and getting to hear all those fabulous performances. I was everywhere . . . at the Kennedy Center, Warner Theater, and then I got to be the audio engineer for the Redskins and then DC United Soccer at the RFK stadium for 20 years . . . and I absolutely adored doing that. And then my involvement with women's music; it was . . . absolutely one of the most amazing times that I could have experienced . . . and I'm so blessed to be a part of that – being at Michigan year after year – all those festivals, meeting all those women. The other part that I got into in Washington, D.C. was I became the audio engineer for many of the different ethnic communities. . . . I absolutely loved listening to all the different music and learning about all the different cultures. That's what led me to ethnomusicology when I decided finally to end my sound company. That led me to my career today, which is teaching at the University of Maryland.

Christine commented about her work with feminist artists at a longstanding and well-known folk music club, performed alongside her academic career:

> . . . it's been a way I felt I could contribute to the women's music movement, and that has been very important to me, to be a part of that in whatever small way I've been able to do that. . . . so my favorite people to do at the X are all those people. I do almost all of the women's shows at the X, partly because those are people that I worked with my whole sound engineer life. I pick up a few new artists here and there – Ellis, GirlyMan,

new people that I'm starting to establish relationships with. But . . . just being a part of it has been so powerful and such a wonderful opportunity that I feel like I've broadened from that.

Certainly the personal relationships that I've established with people over the years and have been able to maintain just because of what the women's music community is, have been very strong, and some of the strongest, most powerful relationships that I've had, even though they're not people that I see on a day-to-day basis . . . but [I have seen] them in an intense way over the past 30 years.

She has also gained a great deal from engineering the music events at the folk music club that are not of a "women's music" genre:

I love music and it's a way that I can help create musical moments . . . I love being a part of that from the inside. . . . I feel that I've developed skill over the years that I can contribute, and it is a way of paying something back to the universe at large, and a lot of that grew out of the women's music movement, but also it's part of a larger way that I can participate in something that's meaningful to me that I feel has the potential for being meaningful to other people, cause I think there's a lot about live music that creates experiences that are important and becoming more important as people become more isolated. So, being a part of that has meant a lot to me. Plus, I love learning . . . , and there's always new stuff to learn and play with.

Her satisfaction occurs despite the sexism that continues to emerge when she works alongside a male colleague at the folk club:

if there are two of us there, a man and a woman, and currently I'm the only female engineer at the X, visiting people will inevitably ask the man, and not me. And I tried to explain this to a really fantastic [man] – one of the most gender-neutral people I've ever met in my life in terms of just how he deals with people – I tried to explain this phenomenon to him, and he said 'Oh no they don't.' Shortly after that we had that experience . . . he was recording and I was actually running the show, and people afterwards kept coming up and telling him what a good job he had done, and he would say 'no, it was her', and they would look at me and they would look back at him, and they would either say nothing, or repeat it to him. . . . This happened several times. I . . . said [to him]: 'That's what I was talking about' and he said 'I get it now.'

What has made each of these women successful audio engineers has been not merely "good luck and grace and skill" as Liz put it, but also an appreciation for art – the art of music and the art of the music production process. Their success in a profession that was and remains male-dominated is a testament to their talent, hard work, perseverance, and high standards. I realized after the interviews that much of the women's music (and some folk music) I have most enjoyed over the years at live concerts and on recordings was engineered by the women interviewed for this chapter. Having learned about their individual trials and successes, I appreciate their work all the more. They are indeed trailblazers and role models – whose notable achievements have been fueled by a passion for music, personal strength of character, and a generosity of spirit in wanting to "give back" to the women's music movement. They have given "voice" to talented musicians and to political causes. It is my great honor to now project their voices to a wider audience.

Conclusion

The women audio engineers studied were trailblazers by working hard to master complex electronic technologies in a field that was and continues to be male-dominated. Although most of them came from families in which parents played musical instruments, that was not a defining factor in their choice of profession. Instead, each of them developed their own love of music, exhibited inherent curiosity about technology, and were for the most part swept up in a current of feminist social change activity that promoted women's musical performance and production.

All of the women encountered gender-based roadblocks along the way – in their own formal education, in the recording studio, or at the concert/rally venue. But most also found male mentors who respected their willingness to work hard and operate with high-quality standards. Women were inherently well suited to audio engineering because they recognized the importance of the human as well as the technical component of their work. While much of their knowledge was acquired informally through on-the-job observation and interaction with other employees, most of the women also took formal courses or attended workshops to add a theoretical component to the knowledge. Each of the women rated their confidence as high, and their accomplishments in the field validate their self-assessments.

Finally, each of these accomplished women exhibited humility and a willingness to share their knowledge with other women entering the field. The latter is a hallmark of the true meaning of feminism. As the talented songwriter and singer Holly Near would later write in her eloquent way, women's music changed lives, and women's lives changed the music (Erenrich and Wergin 2017, 162). The audio engineers described herein have brought that wonderful music into our homes, dances, and concerts halls, enriching us in the process.

Notes

1 I had the privilege of seeing Ms. Cotten perform twice in the 1970s – once at a folk music festival in upstate New York and again at a small coffee house in Brattleboro, Vermont.
2 I was in attendance at that festival and witnessed the described event.

4 Woman in underground Detroit

The non-traditional early occupation of a university graduate

Women in the 1970s and 1980s, buoyed by the feminist movement and supported by federal regulations, made concerted efforts to enter male-dominated skilled trades professions. Jobs in the trades offered portability and higher salaries than those paid to women in traditional service, manufacturing, and certain white-collar occupations – even in fields like teaching and nursing. In 1978, President Carter's Executive Order 11246 and Department of Labor regulations expanded upon President Johnson's earlier National Apprenticeship Act and Executive Order 10925 by requiring increases in female apprenticeships and hiring on federally funded construction projects (Office of the Federal Register 2016; Frank 2001). With success on the regulatory front, feminist activity took the form of support groups for women seeking to enter apprenticeship programs. Examples of such were: Wider Opportunities for Women (2017) in 1964, Nontraditional Employment for Women (2017) in 1978, Tradeswomen, Inc. (2017) in 1979, Chicago Women in Trades (2017) in 1981, and Step Up for Women – a building trades pre-apprenticeship program established in 1979 by Ronnie Sandler and "sponsored by the Michigan Women's Commission in Cooperation with the Michigan Department of Labor's Office of Women and Work" headed by Hilda Patricia Curran, in cooperation with Lansing Community College (Step Up for Women 1979; Michigan Women's Hall of Fame 1998; LaTour 2014). The Coalition of Labor Union Women (CLUW) was founded in 1974 to advance women's equity in the workplace and in unions, with the following stated goals: "to promote affirmative action in the workplace; to strengthen the role of women in unions; to organize the unorganized women; and to increase the involvement of women in the political and legislative process" (CLUW 2019).

This case study focuses on one woman who acquired a most unusual non-traditional job in 1979 given her undergraduate degree in Special Education. Unlike other women of her time who pursued traditional skilled trades jobs in the building and manufacturing sectors, the subject of this chapter worked as a Water Utility Worker in Maintenance and Repair for the City of Detroit Water and Sewerage Department doing sewer repair and later got an apprenticeship to become a water treatment plant operator. Most of the information in this chapter came from several interviews I conducted with her in 2013 prior to her untimely death from ovarian cancer in 2014; other details are based on personal knowledge gleaned over the 45 years that I knew her.

Portrait of a pathbreaker

Norma Jean Lim Andres was born on September 27, 1949 to immigrant parents from the Philippines: Roman and Pacita Andres. Her father had been a chef at the Waldorf Astoria Hotel in New York as a young man, and following marriage and the ownership of a small diner in Mt. Clemens, Michigan, he ended his career as the head chef at the Colonial Hotel in the same city. The Colonial housed one of the last mineral baths in a city that had been a mineral bath destination for the wealthy from 1873–1974 (Mount Clemens Downtown Development Authority 2013). Her mother, whose maiden name was Lim, was a stay-at-home mother while the children were young, and then did restaurant work with her husband. When he died in 1973, Pacita worked as a security guard and then as a "blue light special" announcer at K-Mart. Norma's maternal grandfather, Pedro Lim, had come to the Philippines from Canton, now known by its Chinese name of Guangzhou. Spanish names like "Pedro" and "Andres" were the result of Spanish rule of the Philippines, which lasted from 1571 to 1898 (Slack 2014). Norma was the middle child and only girl born to Pacita and Roman, and was raised Roman Catholic in a modest house in Clinton Township, Michigan – approximately 25 miles northeast of Detroit. The Andres were the sole Asian-Pacific Islander family in the neighborhood, but apart from one incident with a neighborhood bully whom Norma unexpectedly laid flat in an impromptu boxing duel called by her older brother, the Andres children enjoyed their childhoods and were academically successful and popular.

Norma was relatively short (5'2" tall in adulthood), but had an athletic build thanks to years of playing sports in school and subsequent participation on women's volleyball and softball teams along with martial arts training. She worked while in high school, starting at age 14 as a bus girl at a local steak house. Her parents encouraged all of their children to get university degrees. Norma recalled:

> It was actually my mother who wanted her children to graduate from college [in] . . . America. . . . They [her father and mother] both saw education as important and [wanted us each] to get a college degree. They knew that already during those times that . . . education can never be taken away from you; also it would mean getting a better job, having a better life. Essentially they were correct on that.
>
> (Andres 2013)

She enjoyed science and theatre during her high school years, and initially chose the technical field of theatre lighting as her undergraduate major. The reason for her choice was: "I loved the production. I liked doing things, being in production. . . . from high school I liked it." Yet when she studied theatre production during her freshman year at Wayne State University, her confidence waned when she realized that other students were "so advanced" in their knowledge, and that her high school education at Chippewa Valley came up "very, very short" compared to that at other schools. Ultimately, she decided to change her major because:

> I didn't have the confidence. When I started looking at the experience that people had already with lighting in their lives, I didn't have the confidence at that time to pursue [it]. I was pretty sheltered; I didn't have a car to do that kind of stuff [theater work]. . . . I went to a commuter school – Wayne State; it wasn't a [residential] campus school where I . . . could go places and learn things.

Wayne State University (WSU) was an urban campus ripe with social change opportunities in the late 1960s and early 1970s. The Viet Nam war was raging, black unionists at workplaces like Dodge Main were rocking the Detroit labor movement, and a major feminist conference held on campus in the fall of 1970 prompted members of a registered student organization to which Norma belonged[1] to change its name from the "Association of Women Students" to the "Women's Liberation Organization" (Wayne State University 1969; 1970). Norma's undergraduate education included her involvement in campus peace/ anti-war and feminist activities, and her joining of the Delta Zeta sorority which had other feminist members. Sororities at WSU's large commuter campus did not have dedicated houses, but rather sponsored social gatherings and other activities as a means of networking.

The courage Norma had demonstrated in childhood when she delivered an upper cut punch to the neighborhood bully who was older and larger manifested itself once again during her first semester of college while working part-time at the Hudson's department store at Eastland – a mall in Harper Woods, Michigan. Mr. J.L. Hudson, Jr. called for a meeting of all employees at the nearby movie theatre at which he spoke against union organizing. Norma had pro-union sympathies, but following the talk, Norma approached Mr. Hudson for another reason – to ask if she could participate in the company's annual Thanksgiving Day parade on Woodward Avenue in downtown Detroit. She was prepared with her name and phone number written out to hand to him. She got a call the next day, and participated in the parade as a walking pirate.

When theatre lighting didn't seem to be a viable path for her, Norma thought about teaching and selected Special Education as her major, because she "liked helping the underdogs" and the work seemed "more challenging" than other aspects of education. That changing of majors tacked extra time on to her undergraduate studies, since certain prerequisite courses had to be taken. After graduating in December, 1975, she couldn't find a teaching job because of a recession and an oversupply of teachers. The national unemployment rate that year was 8.5 per cent overall, and 9.3 per cent for women (Nilsen 1984, 22). In 1976 Norma took a course and worked as an Emergency Medical Technician, drawn to that job by her love of science and medicine. During that period she decided to take a City Civil Service exam in pursuit of a job at a Detroit hospital. Of the 6,000 people who took the exam at Cobo Hall, Norma scored #11, and the top female; the first ten had been medics in Viet Nam and got veteran's preference.

Those test results qualified Norma to work as an Emergency Room Medical Attendant at Detroit General Hospital (DGH) – a job she loved. Soon after completing her undergraduate degree and while working at the hospital, Norma enrolled in a master's degree program at WSU in education focused on learning disability. She completed all of the coursework and an internship at Lafayette Clinic for mental health – describing the latter as a "great experience." She selected a thesis topic on incest and homosexuality with mental illness based on her work at the clinic:

> There was a young fellow there who was homosexual, and he was a nice looking guy, like the stereotypical golden boy next door who could be the homecoming king . . . his father was a coach and wanted him to be in sports, but he had no interest in sports.

Norma didn't feel that the clinic was doing any rehabilitation or listening to him because they wanted to "straighten him out" which she found "very questionable."

The hospital's life span as a public hospital was cut short by an attempt beginning in 1976 "to turn Detroit General Hospital over to the private, then non-profit Detroit Medical Center" – an effort opposed by the hospital's union, the American Federation of State, County and Municipal Employees (AFSCME) Local 457 headed by President Hazel Edwards and by many Detroit residents who recognized DGH and the Dodge Main auto plant as providers of good jobs and benefits to African-American and other workers (Bukowski 2012, 2–3). Layoffs resulted, and Norma was one such affected employee in September, 1978. One casualty of her layoff was the suspension of her studies, since she had to devote her time to searching for other employment, and wasn't able to complete her thesis. The City paid her unemployment benefits and later offered her a job in building maintenance, which she declined. Detroit would have suffered even more job loss had the city, headed by Mayor Coleman Young and with the support of AFSCME Council 77, not won an out-of-court settlement preventing the U.S. Department of Labor from restricting the use of Comprehensive Employment and Training Act (CETA) funds to re-hire some of the employees laid off months before (AFSCME 1976).

Work as a Water Utility Worker

In the fall of 1979, Norma obtained a job with the City of Detroit Water and Sewerage Department as a Water Utility Worker in Maintenance and Repair, work that had been the province of men. She worked at the Northeast Detroit Water Treatment Plant, located near 8 Mile Rd. and Van Dyke. Several other women who had been laid off from the hospital also got jobs at the City Water and Sewerage Department at the central administrative office on Russell St, in Detroit's Eastern Market area. The Maintenance and Repair central office was located there as well. Norma recalled:

> We all saw each other, and we were happy to get a job because it was recession time. . . . So this was real interesting for the men there because this was probably one of the last fields where there were mostly men and 'men were men and only men can do this type of work' and that kind of stuff. So there were some men that were resentful; very few showed their resentment – I think they were actually curious about what us women could do. And many of them found they liked actually working with some of the women because we were a lot of fun. And we listened.

The motivator for Norma to pursue the non-traditional job at the Detroit Water Department was financial ". . . at that time I had little savings and I lived in a studio apartment [in the Indian Village section of Detroit] that was $125/ month that included all utilities except for my phone."

Tasks, tools, and informal learning

There were a number of tasks that Norma performed as a Water Utility Worker. She described them as follows:

> We learned how to dig up broken water mains and repair them. We learned how to dig up broken sewer lines and repair them. We learned . . . the protocols of being assigned to a crew; because there was [sic] all different types of tasks besides digging up and repairing water and sewer lines. We also . . . took out fire hydrants . . . we did other things like turned off water for unpaid accounts. There were hundreds of them. When you add up all the water . . . throughout the years [for which] the Detroit Water Department didn't collect . . . it's just amazing.

Non-payment of water bills was apparently for "all kinds of reasons" – not necessarily poverty: "some people who could pay just didn't want to pay . . . 'cause they still believed that their water's not going to be turned off. . . . It's not only residential; it's also commercial. There were businesses that didn't pay their water bills. . . ."

The maintenance and repair work required Norma to use:

> All different kinds of tools: wrenches, screwdrivers, shovels; I learned how to use a jackhammer . . . when I was assigned to clean out sewer lines we used a power router with different auger bits to try to get things unblocked . . . also we used a jet hose – a big truck that has a hose like a vacuum; we

would use that to suck up water from the streets or catch basins or water gates, or we would use the high-pressure water to unclog sewer lines.

She always worked on teams of two or more to perform such tasks and "was taught by the men on those crews and other crews" how to use the needed tools. When asked about whether men with whom she worked resented having a woman on their crew and whether they held back sharing their knowledge with her, she replied:

> I think that some of the men were apprehensive about teaching women, for whatever reason. Some had said they had never worked with a woman before. 'Gateman Jackson' was the one that knew how to repair water gates; he was the go-to guy for that. . . . I guess he loved working with me, because here we were taking a pump out of the back of a truck and a pump is used to take water out of a manhole that was filled with water so that we can get to the gate, 'cause the gate's leaking. So we'd take the pump out, turn the pump on and get the water of the manhole. He didn't have any teeth, so he and I are carrying this pump out of the back of the truck to place it next to the manhole. He said: 'In all my years, I never thought I'd be working with a woman.' And I said, 'Well, Gateman, there's always a first time for everything, and I'm sure glad I'm working with you.' So we got along really well 'cause he taught me little things, and I think that he liked that, and I loved that he liked it, 'cause he knew a lot of stuff. And he was very happy to teach, show me because I listened to him, and he was very happy to teach and show me 'cause I think that he saw that I respected his knowledge of repairing water gates.

She recalled an example of his instruction to her, imitating his toothless speech as she conveyed the lesson:

> '. . . the first thing you do when you have an order and you look at it . . . and when you go and you take the water out and you look at it again and see some leak, do you turn off all the water in the area and undo the water gate and put it back together? Of course not. All it may need is just a little tightening like you would do at home with your faucet. Just taking a little wrench and maybe doing a little tweak on it, and there's no leak anymore.'

She added: "He had worked there a long time. He also made his own moonshine and would bring it to work and share it with others on the job."

When I asked Norma what percentage of her work involved climbing into sewers, she replied:

> Not a lot, because you have to realize . . . there are different size sewers. There are sewers you can drive a truck through but those aren't the ones we worked on; we worked on pretty much . . . sewers throughout the city.

The big ones were left for the construction crews. Rarely would we go down the big ones like that, but we did go down some of those.

The types of sewers she typically went down were the ones with manhole covers. The work was dangerous:

You'd open up the manhole cover and aerate it out. They're supposed to have . . . gas meters, detectors, but rarely did we have them. We'd just open up the manhole covers and let it air out. So that was really dangerous . . . because actually once you go down there, there's mud right down at the bottom, and there's methane, and you jump right in there . . . and you stir that stuff up, and methane's something that can knock you out in a flat second.

There were in fact people who "got hurt" doing such work, but not anyone who worked with Norma directly.

Earning respect

Norma explained how she earned the respect of Gateman Jackson: "The questions I asked were legitimate and somehow easy enough and also not intimidating . . . [he knew] that I was not questioning his judgment." Such rapport did not come easily with younger workers:

Some of the younger guys were not like that . . . I think they felt that this was not a place for a woman. . . . Some of them did tell me: 'I wouldn't want my sister doing this job.' I'd just tell them that 'I'm not your mother and I'm not your sister – I'm a co-worker here with you, so you don't have to worry about that. I'm here as a person glad to have a job and I find this area fascinating because it's not like any job that I had before.' I wasn't in competition with these guys for strength.

She recognized that the men themselves had physical limitations, too, and pointed that out as a way of teaching them:

I'd ask: 'Can you dig a big hole like Big John over there?' Big John was this big guy . . . he could lift a heavy shovel with clay on it over his shoulder, 5 feet down in a hole. 'I said, can you guys do that?' They [replied] 'Oh no.' 'Well neither can I, but Big John is; he can do it. So is he going around saying "Can you do this, can you do that?"' And they thought about it and said 'no.' And I said 'Each of us has our own limitations. So it has nothing to do with gender for this stuff; it has to do with the person's being able to learn how to dig properly so you don't wrench your back, you don't hurt yourself, you don't hurt anyone else. Safety is paramount in this type of job. People have gotten killed because they were digging before they had

Miss Dig come in to mark up where the utilities are and they struck a gas line. . . . Who's here to play macho out here? Why would you want to be in competition? This is a job where safety is important so you don't kill yourself or hurt yourself or anyone else.'

Norma had used wrenches and shovels previously, "but just like any of the other crew [members], I learned to how to dig a three by five by five hole, [as] safely as possible."

Confidence

Norma revealed that she "felt real confident" using the tools that were required on her job. She attributed her confidence to: "Knowing my limitations and knowing my strengths – what I could do with the tool and not do with the tool." She had demonstrated courage throughout her life, in confronting the neighborhood bully, in sports, and in her political and personal pursuits. All of those factors likely contributed to her ability to use new tools and learn new techniques. She was certainly not exposed during her childhood to women using the types of tools she used on her Water Utility Maintenance job.

I saw women gardening using shovels . . . arts and crafts – more along that line than anything else, because there was not a lot of opportunity when I was growing up that women would be using jackhammers and things of that sort. In fact, people . . . when I tell them I did that, they still couldn't believe I was able to handle that.

Tradeswomen visit the USSR

While employed at the Detroit Water Board, Norma got to know other tradeswomen in the Metro Detroit area. One such person was Eva Caradonna, a carpenter who – like Norma – had worked in the medical field prior to entering the trades. Eva completed a carpentry apprenticeship, and then in 1983 obtained employment at Cadillac Motor Cars thanks to an EEOC settlement (Michigan Oral History Database Project 2004). Through Eva, Norma learned of a planned 3.5-week visit to the USSR in April–May, 1984 as a "North American-Soviet Tradeswomen Exchange for Peace." Norma, Eva, and others from around the country participated as members of the U.S. delegation. They visited Moscow, Kazakhstan, Lithuania, and the Ukraine. While in Kazakhstan, the U. S. women skated on the Olympic ice rink, and were shown a large Olympic pool there. Norma stated that the Soviets "called every profession a trade." Norma, Eva, and others contributed money and bought a briefcase for the Soviet woman who organized the tour and coordinated meetings with the tradeswomen, but they learned that it would have been better to give it to her at the end of the trip, or better still, to pick a flower, write a poem, or draw a picture to show their appreciation.

Norma recalled additional aspects of the trip that were personally meaningful to her. Because she was studying Special Education at WSU, she stopped in an elementary school in Moscow and they gave her a book. She found Lithuania to be a "beautiful" country, and was given a hand-painted Lithuanian egg while there. She learned that strings for instruments like balalaikas come from animal guts. She observed that it felt "safer in Moscow at 1 a.m. than in Detroit any time." In recounting the overall purpose of the exchange, Norma asked the rhetorical question: "How can you talk about peace or understanding one another when you don't have the same definition?"

A personal adventure

In Kazakhstan, Norma took a 20-minute walk up a mountain alone one day to view the source of a waterfall she had seen. The mountain range was the Tian Shan Mountains, across from Mongolia and on the border of China. There she experienced the "cleanest air I had ever breathed in my life." The people there wore embroidered, pointed hats, and lived in wooden houses. Norma noticed a family's house on her walk, and encountered an older, weathered woman who lived in it. The woman asked a question in a language Norma didn't understand. Norma gave the Mongolian Kazakh people she met "little gifts" from the Water Board, and they in turn gave her vegetables. When Norma descended from the mountain and rejoined her group, she asked their translator what the woman had said, and learned that the question had been: "Where are you from?" due to Norma's Asian looks. The next day, Norma hiked back up the mountain to give a t-shirt and pistachios to the woman who had asked her the question the day before. The t-shirt was written in English and Cyrillic stating: "American Soviet Tradeswomen Exchange." Norma stated that she felt compelled to go up the mountain because she felt that "my soul had been there before."

Change of profession

Following her employment at the Water Board, Norma obtained a position in August, 1985 with the State of Michigan Department of Education as a Disability Determination Examiner in Detroit. Her duties included medical determination of social security disability claims, interviewing claimants and families by phone to obtain medical records, assuring that all determinations met the Federal guidelines related to disability under SSI and SSDI, and providing claimants who did not meet the guidelines the opportunity to determine if they were eligible under the vocational guidelines.

By January, 1987 Norma was hired by the Michigan Department of Labor and Economic Growth-Michigan Rehabilitation Services as a Certified Rehabilitation Counselor – a job she held until her death in 2014. Her tasks were to provide services to individuals disabled as a result of mental illness or various physical conditions such as traumatic brain injury, HIV/AIDS, spinal cord injury, chronic kidney disease, learning disabilities, or substance abuse. In order to do so, she had to gather pertinent medical, psychological, social, educational, and vocational information. She was also responsible for referring and coordinating client training with vocational schools, community colleges, and agencies, helping to place them when ready for employment, and following up to assure employment success. While working full-time, Norma earned a Master of Arts degree in Guidance and Counseling at WSU in 1994, thanks in part to graduate work she had done prior to her hospital layoff. That M.A. degree was directly applicable to her work as a vocational rehabilitation counselor. In recognition of her accomplishments in that role, the Michigan Rehabilitation Association named their annual Counselor of the Year award for her in perpetuity.

Norma also took on volunteer activity, heading the Project Hope program in Detroit for HIV aids, and working with the Kidney Foundation. Beyond that, she continued to enjoy sports, yoga, dancing, and spending time with friends and family.

Conclusion

Although the need for economic self-sufficiency as a single woman was a primary motivator for Norma's career and educational persistence, there was something in her character that propelled her forward. She described that as follows:

> One thing I can guarantee you is my curiosity. . . . I've always been interested in so many different things . . . so if I don't learn something here, there's always something else to learn. So the more I learn, the more I feel I don't know. . . . infinite amounts of 'don't know'. . . . it's such a wonderment to have that feeling of 'don't know'.

Her easy-going personality, work ethic, and respect of others broke down barriers with her co-workers, especially with the most experienced man on the crew, Gateman Jackson. That resulted in him teaching her informally on the job, an experience that women working in physically challenging male-dominated workplaces don't necessarily have. She was also extremely bright, calm in the face of crisis, courageous, and physically strong – like the TV character Xena the Warrior Princess whom she admired. Her connection to other feminists in the trades and in sports offered support outside of the workplace. She was a committed social justice activist, who loved life and was in turn loved by people of many backgrounds. All of those traits led to her success in the most unusual occupation in which she found herself after college, and to continued success in her subsequent career.

Note

1 I belonged to the same organization during that period.

5 Reaching for the sky

Women pilots at major commercial airlines

Women have been attracted to aviation since the earliest days of pleasure flying. In 1910, Raymonde de Laroche of France became the first woman in the world to earn a pilot's license; within a year, seven more women would also attain such credentials – three from France, Russian Lydia Zvereva, Melli Beese of Germany, Hilda Hewlett of England, and Harriet Quimby of the United States – each of whom would become the first person to be licensed in their respective countries (Lebow 2002, 1). Bessie Coleman broke ground in 1921 by becoming the first African-American – male or female – to become a licensed pilot, but had to travel to France to do so since U.S. schools were off limits for a person of her race (Freydberg 1994, 3; Hardesty 2008, 6–7).

Early aviatrixes were drawn to flying for a variety of reasons. All of them clearly loved the thrill and intrinsic reward of soaring in the air – demonstrating great courage in so doing, and breaking with conventional social norms. Clearly, they were comfortable with the technology of airplanes. There were financial rewards as well. Matilde Moisant – the second woman to earn a pilot's license in the U.S. following the path of her friend Harriet Quimby – earned $5,000 for a flight performance of 25 minutes over Dallas, Texas in 1912 (Ganson 2014, 3). But their attraction to flying was far beyond monetary. Lebow (2002) nicely characterizes the motivations and qualities of the earliest women pilots:

> Their appearance on the flying field, at first a surprise, came to be accepted in spite of early opposition from male aviators. Like most trailblazers, these pioneer women flyers learned to ignore criticism that women were unsuited for aviation, taunts that women were interested only in catching a man, and dirty tricks that sometimes resulted in crash landings, and focused on their goal. Adventurers at heart, individualists, definitely too large for the pigeonhole for women in that period, the first women flyers were caught up in the excitement aeroplanes generated. They had to fly and experience for themselves the intoxication of flight; they were courageous beyond imagining to step into flimsy machines, with disaster a constant risk. They were assured – most were familiar with wheels and motors – and viewed the aeroplane as a continuation of the new century's progress.
>
> (2)

Once again emphasizing women's courage, Lebow continues:

> Courageous beyond belief, with confidence and eagerness, the early women
> flyers climbed off their bicycles and motorcycles, out of their automobiles,
> into the flimsy contraptions fashioned from bamboo, wire, and fabric that
> first carried people aloft. Seated precariously between the wings of the
> earliest machines, sometimes below the wing, lacking any protection from
> the elements, these intrepid women, by their example, pointed the way for
> women to attempt the unusual. For them it was conquering the skies to
> reach for the horizon.
>
> (3)

Of course not all women of that period had access to motorcycles or auto-
mobiles. Bessie Coleman overcame the economic and racial limitations of her
small-town Texas birthplace through hard work and study (Hardesty 2008, 8).
Her desire to improve her financial situation led her to join her brother in
Chicago, where she worked as a manicurist at a Southside barbershop fre-
quented by "men of the streets . . . and the underworld – men who had the time
and luxury to care for their fingernails" and in the masculine environment of
her workplace, she learned about the field of aviation from men who had
returned from the First World War – including two of her brothers who had
been stationed in France (Freydberg 1994, 69, 71; Bix 2005, 6). When her earn-
ings and tips proved insufficient to fund her desired enrollment at aviation
school, she started her own business – a chili parlor – and sought the advice
and sponsorship of local African-American philanthropists (Freydberg 1994, 72)

Whether flying planes for the sheer pleasure and adventure of it, as exhibi-
tion/stunt flyers like Katherine Stinson and Bessie Coleman, or in support of
military operations, women's accomplishments in the air were noteworthy.
Moisant was said to have flown at higher altitudes than "most male pilots"
(Smithsonian National Air and Space Museum 2015). Coleman was described
by leading French and Dutch aviators as "one of the best flyers they had seen"
(Freydberg 1994, 85). Female pilots made pioneering solo distance flights, such
as Harriet Quimby's crossing of the English Channel in 1911, Englishwoman
Amy Johnson's flight from England to Australia in 1930, and Amelia Earhart's
solo trip across the Atlantic in 1932 (Women in Aviation International 2015).
During her lifetime, Jacqueline Cochran was reported to have set more records
for speed, altitude, and distance than any other pilot – male or female – in
aviation history (Smithsonian National Air and Space Museum 2015); she
would later head the Women Airforce Service Pilots (WASP) organization,
which received no official government recognition for its courageous wartime
service until 1975 (National WASP WWII Museum n.d.).

Despite the accomplishments of early aviatrixes and those who followed
them, and even with the support of peers belonging to the Ninety-Nines – an
international organization of licensed women pilots founded in 1929 to support
and encourage women in aviation (Ninety-Nines 2019) – women were for many

years unable to obtain pilot work at large commercial airlines. Helen Richey was hired as a pilot by Central airlines in 1934, but faced stiff opposition from the airline pilots' union, and resigned when her flights were reduced and she realized that the job was a "publicity sham" (Cochrane 1997, 12–13). It would be nearly 40 years later during the height of the second-wave feminist movement before another woman – Emily Howell Warner – was able to secure a position as First Officer and later Captain for Frontier Airlines – a regional carrier; a year later in 1974 she became the first female member of the Airline Pilots Association – the very organization that had protested Helen Richey's presence in the cockpit (Smithsonian National Air and Space Museum 2015; Douglas 2004). In 1973, Bonnie Tiburzi was the first woman to be hired by a major U.S. airline (American Airlines 2015; McCullough 1980, 47; Tiburzi 1984; Brady 2000, 393).

It took even longer for African-American women to break through the glass ceiling and enter the cockpits of commercial airlines. Jill Brown was hired as a pilot by Texas International Air in 1978, which at the time was a large regional airline, a position from which she resigned a year later to work at Zantop International Airlines, a cargo carrier (Black Past 2017; Douglas, op. cit., 178; Harty 2014). Patrice Clarke, who had a childhood fascination with airplanes, was encouraged by her mother to pursue her passion. After graduating from Embry Riddle Aeronautical University, she served as a pilot for two airlines in her native country of The Bahamas, then in 1988 obtained employment with United Parcel Service (UPS) as a flight engineer. Two years later she was promoted to first officer, and in 1994 to captain at UPS, becoming the first black female captain of a major airline; when she married Ray Glenn Washington – a captain at American Airlines – that same year, they became the first black couple holding those prestigious positions (Douglas op. cit., 216; Airlines for America 2018). Brenda Robinson was also a pathbreaker, serving as the first African-American woman Naval Aviator in U.S. history in 1980, and ten years later made history as one of the first black female pilots at American Airlines (NBC 2017; National Naval Aviation Museum 2018). She worked at American for 17 years, flying Boeing 727, 757, and 767 aircraft, and in 2016 was the first African-American military woman to be elected to the Women in Aviation International Pioneer Hall of Fame (Women in Aviation International 2016; Afro 2018). The Organization of Black Aerospace Professionals (2018) website highlights other African-American women pilots who have achieved distinction and are regarded as pioneers, such as Captain Theresa M. Claiborne and First Officer Beth Powell (Hicks 2018).

Despite these achievements, women remain a fraction of the total number of commercial pilots. By the end of 2018, they comprised only 5.4 per cent of all airline and corporate pilots worldwide and 1.55 per cent of captains (International Society of Women Airline Pilots 2018). In that same year, India had the highest numbers of female airline pilots of any country at 12 per cent (Shah 2018; Morris 2018; Sinha 2018). In the United States, United Airlines had the greatest percentage of female pilots at 7.4, but overall the rate was 5.7 per cent

(International Society of Women Airline Pilots 2018). The Federal Aviation Administration (2017) uses the term "airmen" in the data it collects and publishes, irrespective of the gender of the pilot. The cost of earning the needed credentials remains an obstacle, as does outright sexism and racism (Hornblower 2001). Meyer (2015) has documented a culture of masculinity in private aviation, and airlines were for many years convinced that passengers would not board planes piloted by women (Anderson 2009).

This chapter focuses on female certified pilots working for major commercial passenger airlines in the United States. It is more common to find women piloting military aircraft or smaller commercial jets, and in keeping with the theme of this volume, a case study of women pilots using complex technology in a field of work in which they are still an overwhelming minority contributes to our understanding of the factors accounting for their success. In-depth telephone interviews were conducted in 2014 with five female pilots working for three different major U.S. commercial passenger airlines – Janet Patton, Terry Siciliano, Erin (surname withheld), "Emiko" (pseudonym), and Juliet Lindrooth. Their work experience as pilots at the time of the interviews ranged from 10 to 27 years, with an average of 17 years. All but one held the rank of Captain at the time of the interviews or earlier in their careers at smaller airlines.

Motivating factors for pilots studied

Passion for flying

Like the aviatrixes who went before them, each of the five pilots had a passion for flying – in some cases dating back to childhood. Janet Patton's family lived near an Air National Guard base, and she recalled that she "developed a love of aviation and flying at a very young age" and resonated to the sound of airplanes as early as age 6. By the time she was 14 she joined the Civil Air Patrol – a non-profit auxiliary of the U.S. Air Force engaged in aerospace education, cadet programs, and emergency services. Through that organization's newsletter she learned about a scholarship offered by the Michigan chapter of the Ninety-Nines, and earning it allowed her to attend a week-long flying school at age 16, culminating in her first solo flight. Working at a nearby airport while in high school enabled her to take additional flying lessons and ultimately to earn a private pilot certificate.

After high school she enrolled at the only public university in her state that had a four-year aviation degree at that time. Hoping to begin her flying class "right away" she encountered:

> a lot of discrimination from the head of the program. And when I didn't get a flying course when I met the prerequisites and he allowed a man who didn't have the prerequisites into the same course that I needed in order to continue flying on time, I decided that that school was probably not the place for me. So I ended up looking at my other options . . .

It became even more evident that she was being hindered when she met with the head of the program and told him she really needed to get into the course in order to finish her degree on time and he replied: "Why do you want to be a pilot anyway?" Without hesitation, Janet asked: "Why do *you* want to be a pilot?" and recalled: "I realized at that moment that that was not the university for me. If the head of the program, who is supposed to be promoting the program asked that question, I really didn't belong there."

Janet transferred her credits to a community college, moved back to her parents' house, and built up her flying time. The two-year degree she subsequently earned in Aviation Flight Technology enabled her to complete a Bachelor of Science in Aviation Management at a different university in 1992. However, that accomplishment coincided with a tight job market. "When I got my degree . . . the aviation industry was really bleak. A lot of the airlines [in all sectors of aviation] had furloughed pilots." Janet therefore returned to flight instruction in order to build up her time, and used her Aviation Management degree to get a job working in security on the midnight shift at a large metropolitan airport. She did that for a year, continuing to apply for jobs at all of the commuter airlines in the area, but was not called for interviews.

Her "persistence paid off" when she obtained a pilot job at a large cargo company. "They had done really well, so by the end of '93 they were starting to hire again," and she was "very fortunate" to get into the first groups of pilots that they hired. "From there my career just really took off." She flew a four-engine turbo propeller cargo plane for 2.5 years and got "really good experience" which helped her to obtain a job that proved to be her path to becoming the first woman pilot for a national passenger airline.

Janet's hiring there was encouraged by a female safety inspector at the Federal Aviation Administration (FAA) with principal operations oversight for that airline, which had no women pilots. The two women had met through their involvement in the state chapter of the Ninety-Nines. Janet explained the ways in which that inspector supported the recruitment of women. She had to approve revisions in manuals, and would change pronouns referring to pilots from "he" to "he/she" in "anticipation of a woman pilot being hired." She was "instrumental" in Janet's getting an interview at the airline, resulting in her being hired in April, 1996. She advanced to Captain in about 1.5 years, and became the first woman to hold that position in the company. By the time she left that airline for a larger, major commercial carrier, they had hired enough women to constitute 10 per cent of their pilot workforce in less than 2.5 years, which was "absolutely unheard of for any other airline." Ultimately, though, recognizing the many work years ahead of her, she decided to look for work at "a major airline that had a history of being around longer than" the smaller one at which she was employed.

She applied to two major carriers, and was hired by one of them, "which was very fortunate because I got hired in the first wave of [new] hiring" in 1998. That gave her enough seniority to avoid being furloughed after 9/11. Still, as is the case with all airlines, seniority is company-specific, so she went from being a Captain on a DC 9 at the smaller airline to a Flight Engineer at the larger company flying Boeing 787s. She went on to become First Officer in most of the airplanes at that company and she has opted to remain in that position in order to raise her two girls as a single mother (divorced).

"Emiko" was born and raised in Tokyo, Japan, and lived there until she graduated college. At about age 10, she went on a school field trip to Haneda Airport, and as she watched the planes "going up and down and landing and taking off, I thought that was just amazing – it was so cool." She added:

> I remember my first airplane ride was when I was about 12 with my family on vacation, going to a southern island of Japan. And I just thought that was so cool . . . walking down the jet bridge. I remember the airplane ride was so bumpy, and my father and my brother were just terrified – white-knuckled and looking like they were going to throw up, but I enjoyed it, I really loved it. And I thought . . . 'I want to do this.' . . . Also I wanted to get out of Japan; I wanted to go to different places. . . . but couldn't become one [a pilot] because of gender.

Her personality was also a factor in her gravitating toward a non-traditional occupation.

> I was always . . . a class leader . . . [and] the leader person among my friends . . . it's just my personality. . . . According to my mother, it's not very Japanese woman-like, but that's who I am. . . . I never thought about 'that's not a woman's job.' Growing up I always had more male friends than female. I'd climb the trees or go for bike rides. I was always a tomboy. . . . I never thought about becoming a flight attendant.

Yet her ambition to be a pilot was thwarted initially:

> . . . when I graduated from college in Japan, I thought about becoming an airline pilot. I just wanted to do it. But back in 1992 when I graduated from college . . . no Japanese airlines were hiring females . . . , so I couldn't do that. Then I thought . . . 'ok, maybe I can become a flight attendant' . . . [but] no – I was too short; Asian airlines had very strict standards – height, weight, hairstyles, everything; I was maybe just ½ inch too short. So, that was it.

She decided to attend graduate school in the U.S., earning a master's degree in Intercultural Relations from Lesley University in Boston. Emiko spent two years looking for full-time work with benefits in that field, and in the meantime worked three jobs to support herself – one at a university helping international students get co-op internship work, a second teaching Japanese, and a third waitressing at a Japanese restaurant. One of the restaurant customers she served – also Japanese – was a male flight attendant who told her about his work. She then decided to become a flight attendant for "Nation Airlines" starting in June of 1998 – work that she did for five years. In the course of that occupation, she "saw female pilots flying . . . and I said 'Oh sure, I can become a pilot.' So that was it; I thought about it, now I see the people flying, and I said: 'I'm going to become a pilot.'" There were five to six female pilots working there and "Every time I flew with those female pilots I asked: 'how did you do it?'"

Emiko began to take flight lessons in 2000, and during that period worked as a management/administrative intern to coordinate daily operations of the flight school and assist with other duties in operations, sales, and mechanics. By 2002, she worked as a flight instructor while still employed as a flight attendant, teaching students how to fly a "very basic" single-engine piston propeller airplane with six round analog instrument gauges known in the industry as a "6-pack" – the very type of plane on which she had learned to fly. The next plane she flew was a twin-engine propeller plane with the same type of instrument panel. She left flight attendant work behind forever in 2004 when she was hired as a First Officer and then upgraded to Captain (pilot in command) in 2011 at a regional carrier for a major airline, work that she did for nearly ten years before transitioning to become a First Officer pilot at a different, major passenger airline. The regional airline used an Embraer 145 jet plane, which carried 50 passengers and used a digital screen monitor or "glass cockpit" instead of the round gauges – "a very nice" improvement and had autopilot technology. Once she joined the major airline she began flying Boeing 737s.

Like Janet and Emiko, Juliet was intrigued by flying very early in her childhood:

> When I was 3 years old, I was watching a TV program where a guy was flying this jet low-level across the ground – we call this contour flying . . . he was an Air Force fighter pilot. And I . . . thought that was just fabulous. I looked at the TV show . . . and I thought 'I want to learn to do that' . . . and every year at Halloween I would dress up as a pilot. Every year my parents would ask me at Christmastime 'What do you want' and I said 'I want to learn to fly an airplane.' I just never stopped. I bugged them, and bugged them, and bugged them about flying airplanes. They got me a horse in [the] hope that would calm me down in my need for speed, and so I did barrel racing with my horse because I liked the speed. And I liked jumping my horse because I liked being up in the air. . . .

Juliet said that her father was an airline pilot, who made a lot of money and travelled. When she was in grade school, and he would go to Hawaii:

> . . . he'd always come back with a sun tan . . . and then he took us to Hawaii as a family. And I thought 'I want to go to Hawaii – how do I do this?' . . . my parents said 'if you want to travel, you should be a flight attendant' and I said 'No – I want to be flying the plane; I don't want to be a flight attendant!' And my dad actually never told me that there weren't

female airline pilots . . . and so when I was 15 he finally said 'Do you still want to learn to fly?' And I said 'Please, please – how do I learn?' I thought I had to join the military like he did. So I thought I was going to have to wait until after high school to learn how to fly. And he said 'No, we can start now.' . . . I was just beside myself – I was so happy in the air that I never looked back.

As a result of those flying lessons at a local airport, she soloed on her 16th birthday – the youngest age permitted for solo flying – and earned her private pilot's license at age 17. Shortly thereafter and before high-school graduation, Juliet received an instrument rating, meaning in her words that one can "fly in the clouds." She went away to college, culminating in a degree in business administration with an emphasis on marketing, and a minor in psychology. She "really loved history" and "excelled" in it, adding: "I would have been a history major if I thought there was a chance that I could ever make a living being an historian." As she considered working in aviation, she thought it important to have a back-up profession, because pilots must take a physical exam every six months, "and if anything goes wrong on that physical you lose your career."

Juliet worked two full-time jobs in order to fund her education, which didn't leave time to take flying courses. She explained:

> To do this career, you have to be very, very motivated and be willing to sacrifice. While my friends were going to spring break – I didn't even know what spring break was actually until after I graduated – spring break to me was working extra shifts on my jobs so I could make more money.

After college Juliet went to work for Xerox in sales in order to pay for flying courses and ratings: 1) commercial – which allows one to get paid to fly; 2) multi-engine; and 3) flight instructor, which permits the recipient to teach people to fly. She quit her "very lucrative job at Xerox" in 1990 to give flying lessons which she did for five years "because the airlines weren't hiring and they required a lot of experience to get hired – even at the regional . . . small commuter level." She ultimately obtained work flying for a regional airline, and was upgraded to Captain within a year. She was hired by a major commercial airline in 1999 as a First Officer – at first flying a McDonnell Douglas 80, and later a Boeing 767 International. Her accomplishments are all the more impressive in that she was a single (divorced) mother of two children when she obtained the latter position, at a starting salary of $24,000 annually.

Erin's love of airplanes also began very early in her childhood. She stated:

> I came from the womb loving airplanes. I'm not sure why. My parents said from a very young age if something flew overhead I stopped and looked up, and nobody else noticed it. I've just always loved them, and I always thought it would be neat to be able to soar above the earth. . . . My friends had Barbie dolls and I had airplanes that carried the Barbie dolls

She had "always wanted to be a pilot" and even knew the type of plane she aspired to fly – a Thunderbird F-15 for the Air Force. Her familiarity with jet fighters no doubt stemmed from her having grown up on military bases due to her father's work in missile maintenance for the Air Force. Yet she lacked female pilot role models so "it just didn't seem like a viable career." Although he encouraged Erin and her younger brother to pursue their dreams irrespective of gender roles, he also cautioned her that being a fighter pilot involved more than the excitement of flying fast planes:

> I had a very wise father who sat me down my senior year of high school and said: 'I think you could get an appointment to the Air Force Academy. I think you would probably make a fantastic pilot, but you're a girl and right now in 1993 girls aren't flying in combat yet. They will, but your chances of being a fighter pilot are really slim. You're very sensitive; you don't like to kill people or things. . . . I think the idea of you being a fighter pilot and having to actually kill people for a job wouldn't make you very happy. And if you're at all considering the fact you might be gay, I would maybe caution you to find a plan B [career] outside of the military.' Again, my dad was very wise. And I'm thankful for that conversation because it helped me to see that probably a life as an officer wouldn't suit me very well. And I'm very thankful that I didn't pursue that.

Interestingly, Erin's early interest in flying did not carry over to the technical aspects of airplanes, or to other technologies. She recalled:

> I was very machine averse – it's a funny story in my family. I knew nothing about cars; I couldn't tell you the difference between pneumatics or hydraulics, nor did I care. My dad and my brothers were always working on cars and I couldn't think of something more dumb than that. I [only] knew to change my oil when my oil light popped on on the dashboard. . . . So when I decided to become a private pilot, my dad just laughed and laughed. He was very supportive, but the irony of that was not lost on him. And so the learning curve for me was straight up, because I had a male teacher both for ground school and for flight instruction, and he would say, 'Well, it's just like the carburetor in your car or it's just like the transmission.' And I'd say – 'I don't even know what a carburetor is!' . . .
>
> There's an expectation that if you're learning to fly an airplane, that you know certain things, and I knew nothing. I couldn't tell you the difference between AC and DC power, nor did I care . . . It didn't occur to me that I was going to have to learn all the airplane systems and know how to troubleshoot them . . . I just wanted to be in the sky.

Her other occupational interest was veterinary medicine which stemmed from her fondness for animals, but while she enjoyed anatomy and physiology in college, she gravitated more toward the humanities and social sciences than math and chemistry. Ultimately she realized that she did not want to become a practicing veterinarian or a veterinary science researcher, so her love for animals led her to work at a spay and neuter center. She then moved into human medicine and worked for five years in autotransfusion in a hospital setting. Although she "really enjoyed" that work, she saw it as a dead-end career.

Erin's dream of flying lay dormant until a fortuitous event took place. While playing on a coed informal pickleball league, she noticed that one of her teammates – a slight woman whom Erin described as "a stay-at-home-mom Southern belle" named Maureen – always showed up wearing an airplane pin on her outfits. Erin assumed that Maureen's husband was a pilot, and when she inquired, Maureen replied in the affirmative but added that she was a pilot as well, and owned a small airplane. Erin expressed her interest in learning to fly and Maureen said that perhaps one day she would take her flying. About a week later, Maureen phoned to ask Erin if she wanted to go up in a plane with her one evening when she needed to take aerial photos of Mt. Baker in the winter moonlight. She stated that she needed help moving the airplane out of the hangar as there was ice in front of the door, and her husband wasn't available. Erin was "very excited and very nervous." After that flight, with Maureen's encouragement, Erin thought that she, too, could learn to fly.

Within two weeks, at age 27, Erin signed up for ground school and found a flight instructor who would take her up every day and sometimes twice a day, while she continued to work full-time at the hospital. She realized that she had a knack for flying, and for the first time thought about pursuing it as a profession. "I was completely naïve and uneducated. I didn't know anything about the industry; I had no idea what it would take to fly as a profession." So when she got home from her first flying lesson, she Googled "women pilots" and found some "fantastic resources" pertaining to networking and mentoring. Maureen, who had at that point ten years of flying experience, coached her through studies, and served as a sort of a "surrogate mom." Erin got her certificate "really fast" – in about a month – noting "once I make up my mind to do something, I get kind of blind to everything else in life."

Once she recognized that she could make a living as a commercial pilot, she sold everything and "drove clear across the country from Seattle to Florida" to spend a year at an intensive airline training academy. Shortly thereafter she was hired into a private corporate jet flying position and did that for about a year and a half. That corporate job "didn't mesh with the time off and schedule flexibility" that she wanted, so she took a job at a regional airline near Atlanta, Georgia. A scholarship awarded by the aforementioned Ninety-Nines enabled her to become trained as a Captain for a Boeing 737. She really enjoyed the nine years she spent at the regional airline, upgrading to Captain and flying over the U.S., Canada, and the Caribbean, but longed to move back to the northwest coast and was happy to give up her Captain's job and pay when she was offered a job with an airline located there. She reflected: "It took a lot longer than I thought" – 14 years from her initial flying to her being hired at the large passenger carrier in the Northwest, but added: "The flying has fulfilled every ambition I had professionally."

She continued to fly small planes with her friend Maureen for fun:

> Maureen has been there from the very start, and we've had such fun adventures together. We've flown her airplanes all over the place. We tried many years ago to become the first all-female crew to fly a single engine general aviation airplane to Russia. We got stuck in Nome, Alaska with bad weather that never abated by the time our flight permission expired, so we didn't get to Russia. We could see it across the strait, but they wouldn't let us go. We took the last 2 weeks and we flew all over Alaska. Flying there is very tricky. We got into some situations that would make our mothers unhappy – if they knew.

Erin was raised to believe that she could accomplish her life goals. "I didn't know that there were limits out there . . . when the going got tough in my airline training my mantra became 'Failure's not an option.'" Her mother paid for much of Erin's airline training, but has been "fearful and negative" about her flying. Still, Erin noted that:

> she's very proud of me – she loves to tell people that her daughter's an airline Captain, but I always have support from her after the fact – after the fact that I landed safely and I didn't die, she's happy for me.

Only one of the women interviewed for this case study – interestingly the one with the longest career as a pilot – did not think about flying until after college. Terry Siciliano was raised in a family with four siblings by parents who had careers in the sciences – her mother as a nurse prior to and after her full-time parenting years, and her father as a "fluidics engineer" working as a civilian for the Department of the Army. Perhaps because of that exposure, Terry gravitated toward science and math in college, and "enjoyed science immensely" though initially she did not have a clear curricular focus. Her parents "absolutely encouraged" their children in science and math, and "with all the adventures my dad exposed us to – everything he could possibly expose us to . . . camping, and sailing, and skiing. . . . it was great. They said you can do whatever you want." Terry took "every astronomy course they had" at the university and worked part-time in the planetarium during the summers.

Ultimately she pursued a degree in Interdisciplinary Studies, which allowed her to earn a secondary education certificate in Earth and Space Science. She spent an extra year in college since she didn't know what she wanted to study. Terry student-taught in that final year, but "hated it." There were jobs in the field, as "there's always a shortage of teachers in science and math" but since she disliked teaching she was at a loss as to what to do after college.

During college she engaged in competitive sailboat racing for her school, and served as the lead skipper when her team won the Women's National Sailing

Championship in 1982. She had been in sailing boats "all my life. . . . I grew up with sailing – we travelled up and down the East coast. I have 2 brothers and 2 sisters, and we went everywhere racing boats." Both parents sailed: "My mom was great." They owned a lightning-class boat with a cabin – about 18 feet:

> and we would stuff 7 people in that sucker to go and travel the Chesapeake Bay, and anchor and wade into shore with a tent on our back and our supplies, and just camp out – wherever it looked good, every weekend.

About fitting seven people in an 18' boat she said – "isn't it something? We had a full crib for whomever was the infant at the time; it was crazy. My parents were very adventurous."

Her sailing skill would prove to have relevance to a career she had not yet thought about. With her undergraduate degree behind her, not knowing "where I was headed; what I was gonna do" she spent the summer after college racing large sailboats in the Martha's Vineyard area, and then started bartending at a "really nice Irish saloon kind of place" since she'd had "a lot of restaurant experience." There was a fellow in his mid-40s who would come in daily during happy hour – he had "been a Marine pilot, and he was now in a software company. . . . Just to keep his hand in aviation, he taught small airplanes at the local airport. . . ." They would "talk and talk and talk" and he said:

> 'You would just love this – this is for you' because I used to sail and was very competitive in sailing at college. They're both very similar: sailing and flying – aerodynamics and hydrofoil principles. . . . So I would share these stories about sailing, and he'd say: 'Oh my gosh – you've got to go up with me, you've got to go up with me' he pressed and pressed, so I finally took a couple of lessons and said 'You're right' – I fell in love with it. And that's how it started.

Terry was 23 at the time.

When she first took up flying "for fun," she had no intention – or vision – of being a commercial airline pilot like her uncle who was a Captain at a major airline. Yet when he:

> . . . got wind of the fact that I had gotten a pilot license, he said: 'You have to move down here with me. . . . you are going to be an airline pilot, and I'm going to teach you how.' He was based in Atlanta, so I packed up everything and quit my job and he recertified as an instructor, which he had been many years prior, and started flying with me at the local airport. I got a job with the local airport pumping gas and washing airplanes, and flew as many hours as I could as quickly as I could. I bought a used Cessna for that purpose. It took about 6 months to get all the ratings, and then I started instructing . . . at Dulles Airport.

The sequential ratings that she referred to pertained to small airplanes and were: 1) airplane single engine land; 2) instrument; 3) commercial; 4) instructor; and 5) certified flight instructor – instrument (CFII). She found that much of the science she had learned at school was relevant to both sailing and aviation:

> because correcting for tidal current in sailing is the same as correcting for wind drift in flying. Also lift on a sail is the same as lift on a wing. These similar concepts and others made it easy to transfer the knowledge of hydrofoils to airfoils. I think that's what drew me to flying – they were so similar in so many ways.

By that time, she had decided to become a pilot for a major commercial airline, and her uncle was "very instrumental" in helping her "to make the choices I made and doing what I needed to do to reach that goal . . . to rack up enough hours to apply to the airlines one day." Then she learned about the Air Force Guard and Reserve Program which offered the possibility of pilot training for those who have never been in the military but who have the right credentials – meaning a private pilot's license and a college degree. There are one–two slots per year available on a competitive basis, based in part on the amount of flying time applicants have. Terry described that program as "much less of a commitment" than being on active duty and training to be a pilot, stating that the latter involved service for about nine years in return for the training given. "This program only required 6 months of active duty" after which you were free to work for an airline or elsewhere so long as you stayed "in the Reserves for about 6 years. So that was a great deal."

Terry was selected for the program and was based at McGuire Air Force base in New Jersey. She flew C-141 cargo planes, which "haul troops, bombs . . . equipment all over the world for the Air Force." That plane is not used by the Air Force anymore; it's been replaced "with something more modern." Thus, her path to becoming a major airline pilot was both civilian and military, which she characterized as atypical. Her father had thought Terry would become an astronaut, but then when she was in pilot training for the Air Force, the Challenger space shuttle explosion occurred and that event discouraged her from pursuing such a career.

Terry was hired by a major commercial airline (the company for which she still works) in 1988 as a Flight Engineer on a Boeing 727. She noted that at that time "there was a pilot shortage, there was hiring, and people were advancing much quicker. Everybody who was hired at that time was [started as] a Flight Engineer." A Boeing 727, which she said is no longer in service except at some transport companies, had three crew members in the cockpit: Captain, First Officer, and Flight Engineer. The Flight Engineers sat sideways and were responsible for "the fuel panel, the environmentals, the hydraulics" but didn't fly the plane. In 1990 she moved up to First Officer, or Co-Pilot. That same year the Air Force activated her to fly in Operation Desert Shield and later Desert

Storm. She served in active duty for eight months, then returned to her First Officer position at the airline.

She was eligible to upgrade to Captain in 1995, but opted to defer that opportunity for 13 years to help raise her children with her husband, who was also a commercial pilot.

> A typical pilot advances as quickly as possible . . . my contemporaries [were] engineers for 1–2 years, advanced to First Officer, which meant they were flying in the right seat, and then upgraded to Captain. . . . I was eligible to do so, which most of my contemporaries – men – did. I did not do so because I would have had a more unstable schedule on call, and I started having children. I had two girls – one in 1991 and one in 1993. And because the schedule would have been unstable for them, I remained First Officer on the 727 for a long time, then I went to another airplane, but still a First Officer by my choice, because I could control my schedule and be with my kids and provide a more stable environment for them. . . . That's 13 years deferring the Captain upgrades until the children were older. That's one of the things that a lot of the women do. . . .

When Terry did accept the Captain upgrade in 2008, she served as International Captain of a Boeing 767 and a 757 – planes she had flown previously the prior 15 years. She flew to Europe, Japan, South America, Central America, and Caribbean. She indicated that captains are paid a bit more for international flights, as there's more flight planning involved. In 2014 she transitioned to domestic Captain of a Boeing 737.

Role of mentors and family influences

The paths taken by each of the women in this study toward their pilot careers were varied, but with certain commonalities. In two cases, female mentors played an influential role – for Janet it was an FAA inspector whom she met through the Ninety-Nines, and for Erin it was the female pilot with whom she became acquainted through recreational sports. For the other interviewees, men played an influential role either within families or individuals met through service-sector work. Juliet's father who was a commercial pilot encouraged her stated ambition, and although her mother did not work outside of the home, neither of her parents subscribed to gender role stereotypes. She recalled that her mother easily combined riding motorcycles with her husband and children with sewing. Terry's uncle – a commercial pilot – played a key role in mentoring her once she decided to pursue that career. Erin's father encouraged her and her brother to pursue careers irrespective of gender roles, but he also cautioned her away from a career as a military pilot based on his understanding of who she was as a person.

Neither of Janet's parents were initially keen on her becoming a pilot. Her mother's brother-in-law had been shot down while serving as a reconnaissance pilot during the Viet Nam war. During her middle to high school years (prior to her starting flying), her mother tried to deter her from being a pilot by suggesting first that she be an Air Traffic Controller. Janet rejected that, stating: "they sit in a dark room and look at a screen; I don't want to do that." Her mother then suggested that she be a flight attendant; Janet replied: "They're in the back of the plane; I want to be in the front of the plane."

Televised reports of an Air Florida crash into the Potomac River were a turning point for her at age 12.

> . . . I already had a love of airplanes, and all of a sudden when I saw the airliner go down, I said: 'Bingo. That's what I could do. I could be an airline pilot.' Immediately that's what I decided, and much to my parents' horror – they didn't even like flying, I stuck with it. That was at the age of 12. When I was 14 I said I still wanted to be a pilot, and they said: 'OK, whatever'

Ultimately, her parents encouraged her non-traditional pilot aspirations, even though neither one of them had a great fondness for airplanes:

> My mom instilled in me that I could be independent which has . . . served me well. Being a bit strong willed comes from mom. Dad encouraged me to do what I love, and not to do something [a job] as an obligation. That's a wonderful thing.

Still, Janet noted that both parents worry about her being in the air. When she was 19 and back living with her parents while serving as a flight instructor, one Saturday morning they asked "What time can we expect you home?" Janet replied that she couldn't offer a time, since something "might come up" and didn't want them to worry. She advised them that "no news is good news." They accepted that and it has been an unspoken rule between them since that time. They assume that she's always flying and don't contact her much; they now wait for her to contact them. "They make the assumption I'm always flying; that's probably a good thing."

Irrespective of external influences such as mentoring and encouragement (within the family or by casual acquaintances), all of the women had inner drive and personality traits like independence and love of adventure which were central to their desire to become pilots. That was certainly the case for Emiko. Not only did she exhibit leadership tendencies early in her life, but she also spoke her mind, which her mother didn't appreciate because "I was too independent for a Japanese woman." Her father encouraged that independence: "he pushed me to get a driver's license when I turned 18 . . . and bought me a car, and I drove everywhere in Japan" but her mother was quite traditional in her views and behavior, never learning to drive a car or even ride a bicycle. Observation of women within her extended family also expanded Emiko's view of women's societal possibilities, for her paternal aunts traveled, drove cars, and "were outspoken." They may have been encouraged by their own father who had become a successful entrepreneur by establishing a dealership selling U.S. cars in Japan after the Second World War.

Interestingly, the undergraduate fields pursued by the pilots in this study had nothing to do with their choice of career, with the exception of Janet and to some extent Terry. As stated earlier, Juliet gravitated toward history and did not care for science or math. Ultimately she majored in business administration and minored in psychology, imagining that a career in marketing was more viable than one in history. Erin described herself as "musical and theatrical" and as stated earlier, she was drawn to the social sciences and humanities, though she enjoyed anatomy and physiology, but not math or chemistry.

Emiko had been first in her class in math and was also very good in science until high school. However, there she encountered a poor teacher in physics and a good one in English and excelled in the latter subject, which she would ultimately choose as her college major. Terry loved math and science, especially astronomy, but headed toward being a science teacher until she realized that profession was not for her. It is all the more impressive, then, that those women who completed their undergraduate degrees in non-technical fields felt the confidence to pursue an occupation so reliant on the mastery of complex technology.

Aircraft technology types and complexity

The technology of today's major commercial passenger jets has advanced tremendously from the time that most of the women interviewed began their careers and it also varies by aircraft type. Juliet characterized the technological changes she's experienced and learned about since she began flying in 1979 and prior to that as "amazing." She explained:

> When I started flying, we had what's called round gauges – round dials in the airplane, and everything from the navigation standpoint was very manually done. Thankfully, we had VORs – Very High Frequency Omni-Directional Radio beacons that create airways to fly along, and that was an upgrade from technology they had 20 years prior to that, where you had to . . . listen to Morse Code via A-N airwaves. . . . Thankfully those were decommissioned about 20 years before I learned to fly . . . but they were still on the charts, oddly enough, when I was just beginning. So that was a big advancement in technology, and learning to fly with those [VORs] was a lot easier than an "A" and an "N" so that was fabulous.

Still, the system was not completely automated, as it took:

> . . . a pretty good sense of direction to fly VORs to know where you were going, where you were heading, because we didn't have any moving maps in the airplane. So you had your chart . . . on your lap, and you kind of followed along with your finger to see where you were on the chart, and . . . if

you could see out the window, you would look out the window and identify landmarks on the ground, and circle them on your chart. . . . So you just flew from point to point to point if you were flying by visual reference, or if you were flying by instrument reference, you would fly using . . . VORs from point to point to point from radio beacon that's on the ground, to radio beacon that's on the ground. . . . And that was, I thought, spectacular. . . .

Terry offered a similar perspective on the advancement of automation, and how it differs according to aircraft type, stating that pilot tasks:

used to be done a lot more manually, [for example] flipping switches; many things are more automatic now. The one thing . . . that you need to master . . . more than anything else is the navigation system. Most airplanes are equipped with Global Positioning Systems – GPSs that get their positions from satellites. It's much more accurate than years ago when we had some more ancient systems to rely on. . . . Inputting your route and your course and the winds, and all this stuff goes into a computer and [it] tells the air-craft basically how to stay on track and fly. . . . On the 757, you don't have to see anything to land the plane – the autopilot lands it and you never at any time have to see the runway, which is pretty amazing.

She further explained that in the Boeing 757 and 767, the pilot has a choice of how much automation to use: "It's largely manually done unless you have bad weather and then you rely on the autopilot to fly . . . an approach in really low visibility." She added that autopilot is also used "at cruise altitude."

Erin described differences in the levels of technology of various planes she's flown. The Bombardier CRJ 200, 700, and 900 "self monitors and is able to easily tell the pilot what's happening . . . the CRJ does an excellent job of helping pilots monitor the whole health of the airplane . . . with a quick push of a button." Still, she added, the pilot has to intervene; the plane cannot fly itself. Her most recent plane, the Boeing 737, is "fully automated" insofar as once the plane is flown down the runway and lifted off, autopilot can be turned on as low as 600 feet, and the plane will then "fully fly itself to its destination. . . . It's an extremely sophisticated aircraft – it will control its power, it will control its altitude, it will do it all." She explained that in a poor-weather, low-visibility situation a Category 3 auto land system can be employed, along with auto braking which takes into account visibility, runway conditions like ice, wind, and other factors; however, the pilot must still decide what degree of auto braking to use.

In her free time, Erin flies a 1965 Beechcraft Bonanza plane passed on to her by her friend Maureen. She uses it primarily for charity flying for a group called "Angel Flight" that offers free, non-emergency air transportation for children and adults with medical and other needs. She stated that the Beechcraft has a tiny GPS, but no autopilot, no radar; "there's nothing to help you when the going gets tough, so I like to fly that in good weather conditions."

There is no question that pilots and airlines recognize the value of advanced technologies, especially, for example, when needing to anticipate and track changes in weather affecting flying conditions and schedules (Mouawad 2014, B6; Hodges 1998). As Janet stated: ". . . it's important to understand that automation is not our enemy. There are many times when the automation is very helpful. We need to . . . recognize that the system is constantly changing, and know when we may want to use more automation." Her philosophy is to use technology if it makes flying "smarter and safer."

Juliet gave examples of technologies that in her view have improved the work of pilots, based on my 2014 interview of her:

> In the airline world, the [Boeing] 727 was a very advanced airplane, and in the early '80s, Boeing came out with the 757 . . . that was really the airplane that had the first generation of flight management computers. Up until that time, the only way to control the airplane was either hand fly it, or the 727 had sort of an autopilot where it would track a heading and hold an altitude, and that was it. It wouldn't fly the approaches for you; you had to basically hand fly the approaches . . . [the flight management computer] was such a huge jump in technology that when the 727s went away at some of the airlines, a lot of pilots simply retired because they did not want to learn the new technology; it was too difficult.
>
> I would say GPS was the next biggest jump in technology for the airlines. When you combine that with a flight management system, you've got extreme accuracy in all of your navigation. And then they started putting moving maps in the cockpit, where you could actually see your position. . . . the need for paper maps is slowly going away.
>
> The latest jump in technology, which I absolutely love 'cause I'm a geek, is the IPad. . . . in the past we had to carry a [kit] bag that had 50 lbs worth of books and charts in it. All of it was required. And all of that's been replaced with a 1.5 pound IPad. [A couple of pounds worth of paper charts are still carried.] A lot of pilots were getting . . . back injuries . . . picking up the kit bag. So the companies invested a lot of money into giving us IPads. . . . I feel personally like the IPad is easier to manage in the cockpit than all that paper. A lot of . . . the older guys will disagree with me.

Juliet described herself as a person who "embrace[s] technology." She added that some of the older pilots resist technologies like the IPad, but when she shows them how to use it, they state: "This is fantastic; I didn't know it could do this." Older pilots will ask her how to do certain things in Operations (prior to flying). "It's pilots helping pilots."

Pilot skills required

Each of the pilots interviewed for this study conveyed that the skills required to fly large commercial passenger aircraft are socio-technical in nature, meaning that mastery of technology is only half of the equation, with social skills like communication and teamwork being vital as well. Janet offered a clear and succinct description of the three primary tasks of a pilot: 1) aviate (fly the plane); 2) navigate; 3) communicate. She further explained that some technologies like the flight transponder apply to both communication and navigation. Emiko emphasized: 1) the importance of hand-eye coordination – looking at an instrument, thinking and correcting it with your hand; 2) an ability to multitask – flying while also speaking with crew members, flight attendants, and air traffic control; and 3) great communication skills.

Juliet listed additional skills and personal characteristics needed to be a pilot generally and at a large commercial airline: 1) an understanding of spatial orientation and sense of direction; 2) an ability to learn at a high level (e.g., from rote to understanding to application to correlation); 3) dedication: a willingness to sacrifice a good deal of time with family and friends to study; 4) very good study skills since there are two hours of ground school and two hours of individual study for every hour of flying; 5) an understanding of how to manipulate an airplane, aerodynamics, law, and air space regulations; 6) the ability to communicate on the radio and within the aircraft, especially to listen to what air traffic control is telling you; 7) a willingness to strive for perfection, since pilot errors can result in loss of life; 8) a willingness to embrace new technologies; 9) a willingness to take constructive criticism; and 10) an ability to compartmentalize.

Terry also described the technical and social skill sets needed by commercial airline pilots:

> rapid cross check of instruments; scanning to look at air speed, altitude, vertical speed, your eyes are constantly moving around . . . scanning instruments, and that is what goes to your brain and tells you what to do with your hands and feet;
> . . . the need to think ahead in the face of changing weather conditions; good decision making and a good understanding of your systems and instruments;
> operating the aircraft itself: when a system goes wrong, what we can do to minimize the risk;
> familiarity with legal requirements. For example, you must know when you're fit to fly (not fatigued), and must sign in that you're 'fit for duty';
> Good communication skills are vital – when you're trying to coordinate with your First Officer; when you're communicating with air traffic control and telling them what we need to avoid hazards we see – thunderstorms, wind shear possibility, wet runways. . . . Another dimension of

communication is coordinating with flight attendants in the back; some-times clear, concise communications can be really difficult. For example, a flight attendant can evacuate a plane if there's a fire in the back without the Captain commanding or knowing it, which is very hazardous. Commu-nication with passengers takes finesse, [especially when things aren't going well]. Also communications with crew and outside the aircraft: air traffic control; company communications.

The interviewees and others in the industry have cautioned that overreliance on automation can adversely affect operator skill. A commercial aviation editorial offered the following insight:

> For years now, concern has been growing that airline pilots' basic stick, rudder and energy management skills are becoming weak due to over-reli-ance on automation systems. Pilots have become, in the words of Capt. Warren Vanderburgh of American Airlines' Flight Academy, 'children of the magenta,' dependent upon computers that generate the purple-pink cues on cockpit displays
>
> (Aviation Week Intelligence Network 2013).

"Automation dependency" is a term used in the industry to describe what occurs when "flight crews became preoccupied with managing the automation rather than flying the aircraft" and that has contributed to high-profile crashes (George 2014, 20). According to another observer, "While designers are trying to automate as much as they can, complex interactions of hardware systems and their software end up causing surprising emergencies that the designers never considered . . . and which humans are often ill-equipped to deal with" (Charette 2009, 3).

Juliet described that phenomenon:

> We call [the younger pilots who know how to fly only with automation] 'Children of the magenta line' or 'magentites' because the navigation line on the glass displays are magenta. Sadly, these kids coming up don't know really how to fly airplanes; they're not skilled in what we call basic 'stick and rudder skills;' in other words, 'can you fly the plane without autopilot . . . without any of this technology? Can you get into a little Piper Club and fly it?' And the answer most often is 'no' . . . they've only been taught automation and how to push buttons. And therein lies the rub.

As an example, she mentioned the crash of Air France Flight 447, which involved "an extremely inexperienced pilot" flying a complex Airbus 330 plane. She asserted: "People died because of that combination." An article about the accident based on the data recorder of that flight confirmed Juliet's perception, indicating that while the Captain napped, the co-pilot with the least experience was in command of the plane (with another, slightly more experienced co-pilot beside him) and made serious errors of judgement and data interpretation:

> We now understand that, indeed, AF447 passed into clouds associated with a large system of thunderstorms, its speed sensors became iced over, and the autopilot disengaged. In the ensuing confusion, the pilots lost control of the airplane because they reacted incorrectly to the loss of instrumentation and then seemed unable to comprehend the nature of the problems they had caused. Neither weather nor malfunction doomed AF447, nor a complex chain of error, but a simple but persistent mistake on the part of one of the pilots.
>
> (Wise 2011)

Wise further explained that when the "stall" alarm began to sound in response to maneuvers made by the least experienced co-pilot, both pilots continued "to ignore it" based on their belief that is was "impossible for them to stall" a "fly-by-wire" Airbus in which "control inputs are . . . fed . . . to a computer" which then in turn command "actuators that move the ailerons, rudder, elevator, and flaps" and under normal circumstances do "not allow an aircraft to stall." However, with the autopilot disconnected due to the loss of airspeed data when externally mounted sensors iced over, the plane removed "its own restrictions against stalling." It was also noted that neither of the co-pilots flying the plane while the Captain rested away from the cockpit "had ever received training in how to deal with an unreliable airspeed indicator at cruise altitude, or in flying the airplane by hand under such conditions" (Wise 2011).

A well-known example of the value of being able to fly a technologically complex plane while retaining "stick and rudder" skills was referred to by Juliet – the case of Captain Chesley "Sully" Sullenberg's emergency landing of U.S. Airways Flight 1549 involving an Airbus 320. She noted that he was forced to crash land in the Hudson River, minutes after taking off from New York's La Guardia airport because, "the geese went into the engines; probes were covered in goose guts, and weren't accurately reading the situation and wouldn't give him the power he needed when he tried to boost the engine." Clearly his skill and experience enabled the actions he took. In Captain Sullenberger's own words:

> Even though this was an unanticipated event for which we have never specifically trained, I was confident that I could quickly synthesize a lifetime of training and experience, and adapt it in a new way to solve a problem I had never seen before and get it right the first time, and so that's what I did. In 208 seconds. I wasn't sure at the outset exactly what steps I would take, but we didn't have a lot of ambiguity: I knew what happened. I didn't have to waste time with the 'what happened?' phase. I was able to go right to the 'how do I fix this?' phase.
>
> (Sully 2015)

Pilot training

All of the women pilots I interviewed informed me that pilot training is formal, regulated, and overseen by the FAA. Emiko stated that in order to obtain a job at a major airline in the U.S., the applicant must have earned three licenses or certificates: one to pilot private planes – which includes instrument rating – a commercial pilot certificate, and an airline transport pilot certificate. Every certificate requires a written test, prepared by the FAA. Once written tests are passed, the trainee must demonstrate flying skills in front of an examiner.

Janet emphasized the importance of training pilots not merely in the use of controls, but also in navigation for both automated and manual flying:

> . . . in the old days I learned to navigate using pilotage and dead reckoning (drawing a line from point A to point B). . . . That's the most basic form of navigating. Because of that, I have a very strong ability to know where my plane is. Young pilots now are relying on GPS and IPads to show their position. I had to keep track of my position myself. In some ways, technology has made it safer, and in other ways less safe.

She questioned whether such pilots who are reliant upon IPads could navigate if the screen went blank, arguing that it would be important that they have a sense of where they are and what they need to do. "If they lost their [new] technology, could they fall back on basics?" Juliet expressed a similar view: "We're getting to a phase now where kids coming up don't know how to fly without a flight management computer; they don't know how to navigate."

Researchers have called into question current training and testing practices, particularly in the face of abnormal occurrences. Casner et al.'s (2013; 2014) findings suggest that programs emphasizing rote behavior based on memorization should give way to training methods that sharpen cognitive skills such as navigation, failure recognition, and diagnosis. A similar conclusion was posited much earlier by Lena Mårtensson (1995; 1996) of Sweden's Royal Institute of Technology, who also advocated that pilots be involved in the design of technological systems. Barshi (2015) noted that "Current training programs provide detailed coverage of each of the aircraft's subsystems in isolation" and advocates the more holistic and comprehensive line-oriented flight training (LOFT) simulator approach (p. 220). Still, as Erin pointed out:

> There's only so much training that can be done in simulator. You can learn to fly the heck out of the airplane in any situation; any horrible thing that can happen they can simulate and you can learn to deal with it, but when you're in the real world . . . with high-volume air space, it's a different game. And you only get that experience by being in it. . . .

Terry offered a similar insight about the informal-learning aspect of skill acquisition:

They can't train for every situation; it's all about experience . . . [flying] hours and years; you accumulate all that experience and then you just know what to do when it [an emergency] happens. You're not born with it, but you learn it over time; it's OJT [on-the-job training]. . . . [The airlines] expect you to already be qualified when they hire you; they hire people with high qualifications just because they don't have time to teach you that. All they can teach you is how to fly the airplane at our company. Initial training is only 8 weeks long. Obviously that's not enough time to teach all those things [skills like communication and decision-making]; those are the things that just come with time.

Crew Resource Management (CRM) training, which has been in effect since the 1970s (Federal Aviation Administration 2012), is intended to improve human interaction skills, like "group dynamics, leadership, interpersonal communications and decision-making" so that pilots are using "all available sources – information, equipment and people – to achieve safe and efficient flight operations" (American Psychological Association 2014, 1). Tragedies like the crash of Air France Flight 447 demonstrate just how important inter-pilot communication is, especially in automated cockpits:

Unlike the control yokes of a Boeing jetliner, the side sticks on an Airbus are 'asynchronous'—that is, they move independently. 'If the person in the right seat is pulling back on the joystick, the person in the left seat doesn't feel it,' says Dr. David Esser, a professor of aeronautical science at Embry-Riddle Aeronautical University. 'Their stick doesn't move just because the other one does, unlike the old-fashioned mechanical systems like you find in small planes, where if you turn one, the [other] one turns the same way.' Robert has no idea that, despite their conversation about descending, Bonin has continued to pull back on the side stick.

(Wise 2011, 6)

Wise integrated into his analysis quotations from airline pilot and flight instructor Chris Nutter that further explain the importance of pilot verbal interaction and teamwork:

'The logical thing to do would be to cross-check'—that is, compare the pilot's airspeed indicator with the co-pilot's and with other instrument readings, such as groundspeed, altitude, engine settings, and rate of climb. In such a situation, 'we go through an iterative assessment and evaluation process,' Nutter explains, before engaging in any manipulation of the controls. 'Apparently that didn't happen.'

(7)

The men [in command of the Air France Flight 447] are utterly failing to engage in an important process known as crew resource management, or

CRM. They are failing, essentially, to cooperate. It is not clear to either one of them who is responsible for what, and who is doing what. This is a natural result of having two co-pilots flying the plane. 'When you have a Captain and a First Officer in the cockpit, it's clear who's in charge,' Nutter explains. 'The Captain has command authority. He's legally responsible for the safety of the flight. When you put two First Officers up front, it changes things. You don't have the sort of traditional discipline imposed on the flight deck when you have a Captain.'

(13)

Still, teamwork and communication should be present regardless of rank. Women in this study who deferred advancing to Captain to have greater schedule flexibility in order to engage in child-rearing or who switched airlines and had to begin the climb to Captain anew might well be sitting in the First Officer seat with more flying time than or nearly as much as their Captain. Janet described a situation in which she flew with a "brand new" Captain who did everything quickly and didn't communicate with her, seemingly because he assumed she didn't know as much (despite her extensive flying experience). Gender also seemed to play a role in that assumption, and it's not at all clear that training modules include sexism or racism awareness.

The culture of passenger airlines differs from that of the military, and pilots working in the latter domain may have to adjust to an environment that is focused on teamwork when they hire into the commercial sector. Erin explained that many "military guys" get hired by major airlines because ". . . they're great – they've gone out on carriers, they've flown space shuttles . . . they can accurately drop bombs and fire missiles . . . they're amazing pilots but it doesn't make them good airline pilots yet, and their learning curve is really steep. . . ."

She further described how the skill sets differ between the two sectors:

Let's say the F-18 is flown with only one pilot . . . and they're very proficient – they fly at 50,000 feet and 3 times the speed of sound, and they do it all themselves. We here at the airlines have to work in a crew environment. And the Captain is ultimately responsible for the final safety and authority and decision-making in the airplane. But over the years . . . the crew environment has really changed a lot. It's not a democracy per se, but the First Officer's not just there to do what the Captain tells him or her to do. So there's this crew-working-together environment that we call CRM or crew resource management and there's been a huge push [for it] at all airlines in the US . . . we learn to work together, we learn to be diplomatic, we learn to use the experiences and skills and different perspectives of our crew members. Even though I may be the Captain and you might be my First Officer, it doesn't mean that I know everything, that I don't make mistakes and that you don't have something really valuable to contribute at any moment. These military guys who've flown fighter jets are extremely good aviators, but they're not used to working together as a crew. Now they're coming from being a wing commander in their squadron to being the most junior First Officer at [the airline].

Not only is the crew culture different, but military pilots may not have flown in high-volume hub airports, with "lots of radio chatter and different frequencies."

Juliet further explained that there is a certain level of rigidity taught in the military, traditionally referred to as "command and control" which follows those pilots into the commercial passenger jet cockpit:

> These guys get so set in their ways. . . . my husband's not around so I can say this: men are more resistant to change than women are. Women have to be faster on their feet because we're raising children. Children go off in a million directions, and things happen around the household. I think anthropologically, we have to be more adept at change than men have to be, because . . . kids are not constant, and households are no constant.

She also discussed gender in relation to pilot skill:

> This job [flying] may traditionally be a man's job, but a lot of people agree that it's more suited to women than men, because it doesn't take any brute strength . . . to fly an airplane. It takes multitasking skills, which women are *really* good at . . . women also have a lighter touch than men do, and airplanes like a light touch; they don't like to be gripped and strangled. If you grip and strangle an airplane, it's gonna start bucking and kicking all over the place, like a bronco, so you have to have a light touch. I think women are easier to teach than men are, because they're more receptive. They don't have an ego saying 'I know more than you do, and you're not going to teach me anything.' So I personally think that for the most part, women are more adept at it than men – except for my husband!

Confidence levels

I asked each of the interviewees to rate on a "scale of 1 to 5" how confident they were that they could learn the technology involved in flying at the outset of their careers and now, with "5" as extremely confident, "4" as very confident, "3" as neutral/neither, "2" as somewhat confident, and "1" as not at all confident. Three of the women rated themselves as 4 early in their careers, one as 5 and one as 3. Those who gave the lower ratings said that it had to do with the tight time frame required for mastery of the material during training, which in some cases occurred while working other jobs. Over time, though, all of them said that they became "extremely confident" – a rating of number 5.

Surmounting gender-based prejudice

A question posed to all of the interviewees was "have you encountered obstacles in life or in this current profession that seemed to be related to gender-role stereotypes?" Terry, who has worked as a pilot for the longest of any of the other women in this study, responded:

> Oh yes, from time to time. There's always the guy that thinks women should be in the kitchen barefoot and pregnant. They definitely think they're better than you [but] it's just seniority that puts them in the right seat versus the left seat. It's all seniority based, so it's not merit at all. You're always going to have those guys out there – less so these days, but it was more prevalent years ago.

Not only has she encountered prejudice from other pilots but also from instructors, "especially old timers – they just can't get their minds around women being in the cockpit, but they're few and far between." A friend of hers was asked by a flight instructor who had been around for many years: "Aren't you afraid you're going to chip a nail?" She thinks that comments like that aren't as likely to be uttered now due to fear of lawsuits. She still encounters prejudice from customers, however:

> It's constant . . . especially in Central American countries and South American countries. They are in shock when they see a woman pilot at the controls. It's crazy. I just flew a flight and I had a First Officer who was a female, and that got a lot of attention [with passengers saying] 'I've never seen a woman pilot before.' It still happens; it's crazy. But I had my first all-female crew when I was Captain just last week . . . we have a lot of male flight attendants back there too now. It opens eyes.

Such obstacles did not affect her career choices, however, thanks to her parents who encouraged each of their children to "be whatever we wanted to be. They instilled that in all of us, so it never daunted me in any way. . . . And of course we carry it forward with our children [2 daughters]. Pass it on."

Juliet also responded in the affirmative when asked if she had experienced gender-based prejudice in her career:

> Definitely – when I was trying to build time and build experience . . . and through one of my students I met this guy who ran a corporate charter operation and we became friends and he wanted to take me under his wing and help me get time and fly the airplanes for his charter operation, which was absolutely wonderful . . . but the clients didn't want me there.

She did it once or twice, but the clients said that they weren't comfortable with a female pilot, and that ended her opportunity. When asked how she overcame those biases, she replied: "I just figured that they were really shallow, stupid people." She recognized her value when teaching at a "very busy" flight school. Juliet mainly taught other flight instructors primarily for advanced ratings (e.g., commercial certificate) and was "very productive," working six days/week, 12 hours/day "and my schedule was full and there was a 6–8 week waiting list to get on my schedule" – something that not all of the other instructors experienced. "In addition to that . . . men and women would come to our flight school because we had female flight instructors, [and] they wanted to be taught by a woman, thinking that we were . . . better instructors and more gentle." The male owner of the flight school and the chief pilot – a woman – were "very protective of us women."

To study with Juliet or one of the other female instructors:

> you had to be interviewed by both my bosses to make sure that you wanted me as your flight instructor or one of the other women as your flight instructor, and that you weren't wanting a woman for nefarious reasons. So we were highly regarded and highly sought after in the flight instruction world.

Therefore, when the business charter clients didn't want to fly with her and she needed multi-engine time, which is "very expensive" to come by, she developed a syllabus that got approved by the FAA for continuing education for military veterans, then went to the Marine base at Orange County Airport and gave a presentation to young fighter and helicopter pilots. She advised them on how to get civilian ratings to go to the commercial airlines and encouraged them to take flight instruction with her. She added:

> I had even the military pilots *begging* to get on my schedule, because I was the only multiengine instructor approved by the FAA to do these VA benefits at the flight school. I had these guys waiting 6–8 weeks to get on my schedule and fly with me. And they all said how much they enjoyed flying with me, and I would help them prepare for their interviews for the airlines . . . to have these military guys tell me they really enjoyed flying with me really negated, emphasized the stupidity of these corporate clients who didn't want a female pilot. So I was able to really just dismiss it and let it roll off my back as being stupid, and a wasted opportunity for them and for me.

Despite those experiences, when asked if she saw her pioneering work "as part of a movement for social change by breaking ground for women in the aviation field" Juliet replied: "It didn't occur to me at all. I thought that had already been done." She stated that she "was clueless to gender bias until people pointed it out to me." Later in her early years as a Captain, some passengers refused to get on the airplane because she was a woman. She stated at the time of the interview in 2014 that it was still happening:

> I was walking onto the jet, and I checked in with the agent, and a guy walks up to the podium just as I'm getting on and he says 'Well the other pilot better be a man, or I'm asking for a refund.' Well the Captain was standing right behind him, and she said 'We're going to Miami. You're

welcome to come with us or not – we don't care. But the airline is not going to provide you two new pilots because you don't like women' and she walked onto the jet behind me.

Juliet didn't know whether or not he chose to board the plane.

She cited another incident that had occurred a month and a half before our interview, explaining that on Sundays, she volunteers at a museum flying vintage (pre-WWII) airplanes and flies them for the crowds. Her son had flown to Paris with his mother as co-pilot, and when they returned, they went to the museum and she offered him a ride on a vintage plane that she and her husband, who is also a pilot, own. When they landed, they walked over to the crowd. Juliet was answering questions when someone said to her son:

'That was really nice of you to take your mom on a flight like that.' Then suddenly, my very easy going son said: 'My mom is the pilot, not me.' And the person said: 'Aren't you nervous flying with your mom? He said 'She just flew me to Paris and back . . . of course she can fly this little airplane.' A second person made the same assumption.

Emiko and Erin have each experienced varied reactions from passengers and crew. Emiko said that in her experience both female and gay male flight attendants "love to see female pilots" and some customer reactions are negative while and others are positive. Erin reported:

I get a lot of comments from people in airports – mainly passengers – people are astonished that there's a woman pilot. It's never occurred to them that women could be pilots. Or they've never seen a woman pilot. And the comments tend to be either really negative or really positive, but not a lot of in between. I've actually had passengers refuse to get on when they saw that one of their pilots was a woman. I've had other passengers – men and women both – be really super excited and want their picture with me.

Fortunately, she has not experienced career obstacles due to prejudice:

I've never flown with anyone who's treated me differently because I'm a woman . . . the same thing with being gay . . . I just am who I am, and because I have a pretty likeable personality those things don't become issues . . . I'm a professional, I'm good at my job, I happen to be a girl, and I happen to be gay.

A number of the interviewees, while cognizant of being the only woman in a training class or one of the few female pilots in the country, did not identify themselves overtly as agents of social change, but rather as women who have pursued occupations that they enjoy and excel in. That is different for Janet, who encountered gender-based discrimination directly in her undergraduate education and beyond, and was not affirmed by her former spouse who was also a pilot. She has made it a point to support other women seeking to become pilots and those already in the profession, and sponsors scholarships through the Women in Aviation International organization.

Juliet, too, helps women become pilots in general and also at major commercial airlines through her active involvement in the Ninety-Nines. She mentioned a "great article" written by Judy Birchler, a Ninety-Nine, on her blog about women who fly tail wheel airplanes (www.ladieslovetaildraggers.com/). In that article, Birchler apparently described how when she lands her own plane at an airport and if a man also emerges from the plane – whether or not he's a pilot – all of the technical questions (e.g., about fuel, tie downs, etc.) get directed to him. Juliet related to the article personally, because:

> I experience that at work, I experience that in my general aviation . . . , I experience that everywhere I go. . . . I landed and got out of a Piper Cub one time and this guy who's passed away now, but we became very good friends because of this . . . he looks at me, and he looks in the Cub, and there was no other pilot in there and he goes 'You're a pilot?' And I said 'Yeah' and he says: 'I thought you were a flight attendant.' I get that even in my uniform. I had walked up to the gate at Miami to check in with the [gate] agent and one of the agent's supervisors was there and he said: 'Oh, we're not letting the flight attendants down to the airplane yet' and I said 'And I care because?' and he said 'You can't go down.' And I said 'But I'm not a flight attendant, I'm a pilot.' And he said 'You're the pilot?' And I said 'Yeah; was it the long blond hair that fooled ya?' He worked for the airline – 'You don't recognize the uniform?'

In response to my question about how she exhibits such amazing resilience, confidence, and perseverance in the face of gender stereotyping, if not discrimination, she stated:

> I don't know; I just do. If someone tells me that I'm not capable of doing something, I don't believe them, because I think I'm capable of doing anything that I really, really want to do . . . I just have that innate drive. And it just is there; I think you're born with it . . . Everybody – if they want something bad enough . . . they're gonna succeed at it, but you have to want it . . . and you have want to do the work.

Benefits experienced from their work

The final question posed to the female pilots in this study was "What benefits have you experienced from your work as an airline pilot?" Their responses make it clear that despite obstacles they've encountered in the course of their careers, they find their work extremely rewarding.

Terry stated unambiguously: "I absolutely love my job. It's a wonderful, wonderful career." One of the tangible benefits she mentioned was the work schedule of pilots – three days on and four days off – enabling her to spend time with her family. The same was true for her husband who is also a commercial pilot:

> we were both arranging our schedules so the kids always had a full-time parent, and we never needed a live-in nanny or anything. So that's a huge benefit for this type of schedule – shift work – they [the children] always had a lot of quality time with both mom and dad – that was terrific.

She explained that pilots are able to bid on their monthly schedule based on seniority, which is why she remained First Officer for so long (13 years as detailed earlier in this chapter) instead of seeking to be promoted to Captain. A second benefit she mentioned was freedom to travel and see the world. Her daughters, who were ages 21 and 23 at the time of the interview, have been to "a lot of places" thanks to flying benefits: free flights domestically and a slight service charge for international flights.

Family travel discounts, schedule flexibility, and confidence are benefits that were mentioned by other interviewees as well. Emiko appreciates the fact that she is able to fly her parents back and forth to the U.S. from their home in Tokyo – even in First Class – at a reduced cost. Erin also stated that she and her life partner enjoy being able to "see the world" for free. Erin and Emiko cited personal satisfaction and increased confidence as benefits of their work. In Emiko's words: "Self-satisfaction is the biggest thing for me . . . flying an airplane is a big accomplishment. We have to look back and really think about that." Erin stated: "My self-confidence has increased as a result . . . because I've achieved something that not many people have achieved. There are only 450 women airline Captains in the whole world."

Juliet listed a number of benefits she's experienced as a pilot: "The biggest benefit was that I met my husband that way." She felt that the "most entertaining benefit" is "the ability to shock people" that she meets casually when they realize she is a pilot:

> When I'm flying the little vintage bi-planes I wear a helmet and people don't know who I am. And I pull up and I park the airplanes right next to the viewing stand, and I pull the helmet off and my long blond hair comes tumbling out, and the look of shock on people's faces: 'Oh my God, it's a woman. Wow!'

She indicated that she and her husband bought a summer home near the Museum from which she flies vintage planes – the Eagles Mere Air Museum (www.eaglesmeremuseum.com/) which:

> is dedicated to female pilots from the 1920s that helped pave the way for women like me [like] the 1929 Women's Air Race which is still going on

today. Those women started the Ninety-Nines and those women paved the way for me. They broke the ice; they did all the hard work. The WWII Wasps – they did all the hard work. I just got to ride on their coattails – I didn't do anything; I just learned to fly an airplane!

She added that another "big benefit" to being an airline pilot is "my schedule is extremely flexible" allowing her to take a good deal of time off – especially in the summer. Currently she can alter her schedule by swapping with other pilots when her "girlfriends" (other female pilots) invite her to join them on a flight trip. Juliet mentioned that she and her husband have four children. He also has flying schedule flexibility, but she feared that schedules might "change for the worse" after an impending airline merger.

Her schedule flexibility also permits her to engage in free-time flying. She participates in a cross-country air race annually – the Air Race Classic, formerly known as the Powder Puff Derby, which during the year of our interview (2014) was planned to go from Concord, California to York, Pennsylvania – "2400 miles racing little air planes across the country." Her male boss at the major airline at which she is employed has supported her participation in such by making schedule adjustments when necessary. In the next year (2015), she planned to "hop on a plane . . . to Munich, Germany with all my Ninety-Nine girlfriends for an international convention" and was quite certain that she would be able to get the needed time off of work to do so.

Like all of the other pilots interviewed, Janet cited travel as a benefit of her job: ". . . I get to travel and see a lot of things" – mostly airports and terminals and hotels, not tourist sites, but "it's not uncommon for me to be in 3 different major cities around the country in 1 day." She added: "I enjoy somewhat being able to live out of my suitcase for a short time and knowing that I have everything I need there in a small area. I can go somewhere and relax . . . after my time at work."

Both Juliet and Janet described the joy of flying and the view from the air as an intrinsic benefit of being a pilot. In Juliet's words:

I enjoy flying airplanes – I really like it. I do it on my days off. . . . Sometimes I just want to be . . . in the airplane by myself. I enjoy being one with the airplane, I enjoy being up in the sky looking down on the earth . . . the earth is an amazing place from an airplane if you're low level. You see things that you've never seen before. And it's a fantastic way to explore; it's a fantastic way to see the world. It's a totally different view. And I got that view . . . when I was 13 – my dad took me up in a helicopter . . . and my dad's friend took me up in a [Cesna] 172 . . . and we climbed up and I could see all the trees . . . and to feel like you're just up there floating in the air looking at all this stuff at 13 was . . . amazing to me. And I wanted to be able to do that on my own . . . If you do that, especially at a young age . . . you just want it. That was the motivation: to see the world – the entire world. It's really boring crossing the North Atlantic, but if you do that on a clear day you can see the icebergs floating in the ocean. That's something nobody gets to see − even the passengers can't see that. . . . I have the greatest view of any office in the world. I get to see sunrises and sunsets.

Janet offered additional examples of the pleasure gleaned from the visual aspect of flying:

It's very hard to describe the joy that you get from flying and being up and seeing the earth in such an unusual way. There are sights that I have seen from the sky that will *never* be duplicated from the ground − at all. In fact, there are some sights that I've seen from the sky that I've seen once in my whole flying career and I will probably never see again. So having moments like that to enjoy and appreciate . . . is probably one of the biggest benefits.

I asked her to give an example of something she has seen that has impressed her in that way, and she gave three. The first was from early in her career, while flying cargo from Michigan to somewhere in the Midwest, and:

. . . the sun was going down at same time to the West that moon was coming up to the East . . . and it looked like they were perfectly lined up, with us in the middle. For an instant, they were exactly off our wings and in line with us, which was amazing.

The second was a takeoff at about 6:00 a.m. from Toronto to the southwest, about four years prior to our interview.

. . . the morning sun was streaming into my Captain's window in such a way that I realized we would have a shadow when we took off. As we started lifting off I looked to my right and saw our shadow still attached to us, and then separate as I was climbing. Then when I called for the landing gear [to go] up, I watched the gear come up in the shadow. That was pretty cool; normally we don't get to actually see our gear come up. . . . we only see the digital presentations in the cockpit.

The third was on a late-night flight from Toronto to Los Angeles, with her sitting in the First Officer seat:

On our taxi out, the sun was going down, and by the time we took off, the sun had already set below the horizon . . . but as we started flying toward the west, toward LA, the sun . . . appeared to rise . . . and then we saw the sun go back down, so I basically saw two sunsets.

I suggested to her that what she described sounded almost like a spiritual experience. She then drew an analogy between flying and spirituality dating back to an early solo flight, though she was clear to draw a distinction between spirituality and religion: "I felt God's presence on my first solo . . . I felt there was divine presence sitting next to me; I did not feel alone solo . . . I . . . felt at peace – [and] that I was doing what I was meant to be doing."

Conclusion

The five pilots interviewed for this chapter were motivated by an interest in airplanes and a passion for flying, which emerged as early as childhood for most of them. The woman who had not thought about flying until after college found that her well-developed sailing skills transferred from sea to air, and at age 23 she, too, "fell in love" with flying. Mentors and family members were influential in encouraging some of the women to pursue careers as pilots or to persist in their chosen profession. Networking with other women pilots through organizations like the Ninety-Nines has also been important – not only for mutual support but also to serve as role models and mentors to younger women.

Their mastery of complex technology came through formal, highly regulated training and intensive study. Yet on-the-job experience helped them to hone needed socio-technical skills like communication and Crew Resource Management as well as navigational skills that go beyond aircraft control. Interviewees spoke of an appreciation for technology and automation, while also recognizing its limitations, especially in the face of technical failures and other abnormal occurrences.

All of the pilots have succeeded in their fields due to continuing study and the resulting sacrifice of leisure time, finely developed technical and social skills, a love of flying, and sheer determination in the face of gender-based obstacles in their paths. They have taken up the mantle laid down by early aviatrixes, and by their example and mentoring are inspiring a new generation of women to take to the air and "reach for the sky."

6 Toward a women-and-technology paradigm of empowerment

Lessons learned from the case studies

A series of research questions, posed in the Introduction to this volume, served as the basis of interview questions for all of the case studies. There were commonalities in responses, and a few differences as well. In all cases, though, the lessons learned from the case studies point to a way forward for women's use of technology in empowered ways individually and collectively.

Motivation to engage with technology in the masculine sphere

Each of the women in this study had some internal drive that enabled them to push past gender-based cultural obstacles to master technologies traditionally associated with men. The Egyptian women were well-educated, feminist in orientation, and saw cyberactivism as a means to societal equality and democracy. The Newfoundland dragon boat builders were ready to move past the debilitating breast cancer illnesses they had experienced and work collectively with other survivors to learn to use new tools and build something of beauty that they might paddle. The audio engineers were drawn to their chosen profession by a love of music, an interest in technology, and a desire to break barriers in a field dominated by men. The Detroit Water Department Worker gravitated toward science and the technical aspects of theatre while in high school, and her early athleticism contributed to her willingness to tackle difficult physical tasks. Most of the pilots were intrigued by airplanes and flight at young ages, and the one whose interest came in young adulthood was pleased to be able to draw upon navigational skills she had developed as a sailor.

Having childhood encounters with technology supports Stepulevage's contention that "technology is interwoven into everyday life and involves continuous engagement" (Stepulevage 2001, 80), and "a young girl might develop a familiarity with technology as part of everyday living" (63). Janet Patton recalled that as early as age 6 she was attracted to the sound of airplanes flying over her home to and from the nearby Air Force base. Karen Kane remembered her fascination with the mechanics of a tape recorder that her family had acquired when she was about 8 or 9 years old. Three of the Avalon Dragons – Jane

Brown, Donna Howell, and Janice had fathers who encouraged them early in life to help with tasks involving the use of hand tools. Terry's family sailing experience gave her the confidence to take up competitive sailing, and ultimately aircraft navigation.

Families in some cases supported women's internal drive to engage with non-traditional technologies, but that was not the predominant motivating factor. The Egyptian women were all well-educated, and except for the eldest, were exposed to computer technology during childhood in the home and at school. In some cases, one or both parents supported their ambitions, even when cultural expectations were more limiting for girls. The support of family, community, and peers helped breast cancer survivors' involvement with the Avalon dragon boat build once the women had made the decision to get involved with the project. The airline pilots also had supportive parents, family members as with Terry's pilot uncle, or spouses who either actively encouraged their pursuit of flying, or at the very least – as in the case of Janet's mother – turned a blind eye to the potential dangers of that profession.

A number of the audio engineers were exposed to music during childhood through their parents' performances and/or music lessons, but they were not mentored or encouraged to pursue careers in the musical arts. The Detroit Water Department Worker's immigrant parents had no opportunity to attend college themselves, and therefore wanted all of their children to acquire university degrees as means to higher-level careers and a better life. Ironically, Norma's undergraduate degree did not lead her to a white-collar career in teaching – her chosen field of study – immediately after graduation; a career in a different white-collar profession would come some years later, subsequently bolstered by a master's degree in Guidance and Counseling. All of the women studied had the intelligence and drive needed to tackle and persist in their use of the technologies in non-traditional domains.

The "control" and "context" themes discussed in the Introduction to this book emerge in all of the case studies in a blended way, context being cultural and/or political. Each of the Egyptian women broke with traditional gendered expectations early in their lives by their choices of career and/or field of education, influenced in some cases by their awareness of historical feminists in Egypt and political pathbreakers in other countries. Their use of social media tools was propelled by their involvement in feminist causes, the political activism sweeping the country that gave rise to the Revolution of 2011, and their need to connect with other activists locally and globally in the face of government-controlled information dissemination. As Hafa stated:

> It's good to be with like-minded people and to have your ideas vindicated . . . You form . . . really great friendships as well; you create a community of people – a support system is very necessary to survive every day in day-to-day Egypt because sometimes you get very down about how difficult everything is and how huge the fight is to get the things you believe are right [accomplished].

The Newfoundland dragon boat builders sought to become stronger psychologically and physically in the face of a devastating disease over which they had no control. As discussed in Chapter 1, they were encouraged by family members – some of whom participated in the build with them – and by the community at large. They also came from a self-reliant culture that placed a high premium on quality work, notwithstanding gender-based divisions of labor in some families. As Jane explained: ". . . in our [Newfoundland] heritage we would want a boat that is built the way it ought to be built. . . ."

The audio engineers, as Boden explained in Chapter 3, served as a vital part of the women's music movement which took control over all aspects of music performance and production "to achieve economic control of their lives but also to create a women-identified environment that would be safe and comfortable for all participants." Control and context were intersected in women's music culture, with networks of audio engineers, musicians, and production and distribution companies working interactively and generally supporting one another. Both Boden Sandstrom and Liz observed and were taught by women mixing sound at feminist concerts.

The Detroit water utility worker – Norma – was an independent, self-supporting woman who needed to earn a living following a layoff from a prior job at a city hospital. For her, economic control was a top motivation for taking a job with a decent salary. Contextually, she had been active in campus women's organizations during her undergraduate years, and post-graduation was a part of feminist circles, played women's community sports, and had contact with a group of tradeswomen in Detroit. It was the latter group that organized the trip to the USSR in which she participated and found so meaningful.

Several of the commercial airline pilots recognized that being in the pilot seat would afford them a level of control they would not have had in a different aviation job. Janet rejected her mother's suggestions that she seek a career as an air traffic controller or flight attendant. Emiko recalled that as a "tomboy" she had no intention of becoming a flight attendant, but ultimately took such a position as a means to her long-held wish to become a pilot. Contextually, the pilots recognized the legacy of earlier generations of female pilots and benefitted from organizations like the Ninety-Nines, which was formed by those pioneers. Janet's hiring as the first female pilot at a national commercial airline was encouraged by the female FAA safety inspector whom she met through the Ninety-Nines. In certain cases, the women were encouraged and mentored by other pilots – in the case of Erin by a female pilot of smaller private aircraft whom she got to know through a community sports league, and in Terry's case by her uncle who was a commercial airline pilot.

Informal and formal learning

In the Introduction to this volume, knowledge also emerges from the literature review as a key factor distinguishing the victim scenario from one of women's empowered use of technology. In all of the case studies, the women's learned ability to use varied "masculine" technologies served as empowering experiences. As discussed, some of the audio engineers and pilots interviewed reported that they had an inherent fascination with technology, and clearly they and all of others in this study demonstrated technological interest and aptitude. Yet in all cases, technical knowledge had to be learned – through self-study, informal on-the-job learning, formal training, or a combination of such methods. For many of the women in this study, knowledge and skill acquisition was an informal, on-the-job learning process. In other cases, learning was more formal as was the case for the airline pilots, whose training was highly structured and regulated. Even for them, though, the on-the-job experience they gained over time gave them the socio-technical skills to react to unforeseen situations, as Erin and Terry conveyed. The audio engineers and the water department worker were also presented with non-routine conditions on a regular basis. The audio engineers learned primarily by watching and learning from others who did that work – primarily men, and by networking with other feminists who had that knowledge. As stated in Chapter 3, most of them took formal courses or attended workshops to supplement what they learned through informal apprenticeships, out of a desire to learn and understand theory as well as practice.

The dragon boat builders learned through Bruce Whitelaw's verbal explanations followed by hands-on instruction, but that took place in the build workshop and not in a formal classroom. They also learned from one another – a concept known in adult education as peer learning (Le Clus 2011). The Egyptian women learned to use social media tools by trial and error, and as Dalal described in Chapter 2, progressed from using e-mail and e-discussions primarily for social purposes, to using a broader variety of tools "for the causes you're defending and working on." Norma learned some aspects of the sewer maintenance and repair work from the men on the crews who didn't resent working with a woman. The informal learning literature cited in the Introduction to this volume helps to explain theoretically what those women experienced.

Mezirow's theory of transformative learning discussed in the Introduction is also applicable to a number of the case studies. One of the pilots was able to apply navigational skills she had learned as a sailor to flying an airplane. Some of the women described in the dragon boat case study transferred knowledge acquired in the domestic sphere to the traditionally masculine work of building a wooden boat by hand, thereby complementing and expanding existing skills and deriving new meaning from those applications. Those who had not previously used hand or power woodworking tools experienced a transformation in their own prior assumptions about their ability to become proficient in their use. Regardless of prior experience, a number of the women developed strong personal preferences for certain tools and building tasks. Still, as Anne Marie explained in Chapter 1, all of them "were very conscientious about making sure that everybody had a chance to get on all the tools. So it was always about watching out for the other woman to make sure that nobody was being left behind."

As described in Chapter 1, the dragon boat builders emerged from their learning experience as teachers and mentors. They taught members of their families and the community at large that it was possible to recover from the traumatic, scarring experience of breast cancer with new feelings of empower-ment and pride – from their construction of a beautiful boat and the training they underwent to paddle it. They also demonstrated to girls in the community that women could become proficient in the use of woodworking tools and work cooperatively to construct a complex yet beautiful product.

The audio engineers' learning was transformative in that they built upon a familiarity with music – a baseline meaning scheme per Mezirow's theory – and found new meaning in the form of the technical knowledge they acquired and the artistry they applied to their work. Some of the Egyptian feminists – particularly the older ones – learned to use new social media tools to advance their political work, and they too transformed their perspectives on those technologies accordingly. The women in all of the case studies experienced the kind of transformation Mezirow (1994) described by developing new meaning schemes to solve problems.

Technology's limitations

In most of the case studies, the limitations of technology were mentioned. The commercial airline pilots stressed the dangers of overreliance on technology, and the importance of good navigation and communication skills. The audio engineers recognized that the skill in their work involved not only technical know-how, but an ability to interact with musicians and the artistic sense needed to blend sounds into beautiful music.

The Egyptian feminist cyberactivists were clear in their assessment that social media was merely a tool, at best, in efforts to organize people to attend rallies or to stop sexual harassment. As Sara stated, "You can't use Facebook in reaching out to the grassroots, to educate, or spread information, or sensitize, and spread awareness" adding that it takes "getting out and talking to people" face-to-face, especially in neighborhoods. Durar added that face-to-face orga-nizing and political music predated social media use. Dalal's words are also relevant here:

> . . . I don't think the online tools – social media like Facebook and Twitter are really created for deep, sophisticated discussions, but rather for sending short messages . . . for [conveying] sending specific infor-mation – news and to connect with people. . . . If we want to go into deep discussion there are offline places like conferences, and workshops and seminars. . . .

Their experiences bore some resemblance to points raised by Everett (2004) who wrote about the efforts of grassroots black women in Philadelphia who organized a "Million Women March" (MWM) in 1997 with the aid of the Internet and the activist use of ICTs by women globally:

This promise does not mitigate the reality of the counterproductive, hegemonic sexism, classism, racism, and homophobia that dominate the Web and other new-media platforms. While these inroads cannot be minimized, they do not supplant face-to-face activisms and creativity, particularly in the case of the MWM; most significantly they function as phenomenal augmentations. Yes, patriarchy is alive and well online, as in real life. Nonetheless, cyberfeminism and cyberactivism are also alive, well, and destined to grow in instance and influence as women increasingly outpace men in Internet participation.

(1284)

Mojab, too, recognized ICTs as mere tools that may augment, but not supplant needed face-to-face organizing in her study of Kurdish women's use of the Internet:

A new generation of women activists are using the old technology of print, and are fighting gender and national oppressions through feminist journals, posters and leaflets. They also use the traditional methods of oral, face-to-face dialogue in order to conduct their difficult struggle against the four nation-states, oppressive religious traditions, and patriarchal nationalism. While the experience of the Network demonstrates that new technologies can definitely enhance this movement, I contend that no amount of activism in cyberspace can displace let alone overthrow patriarchy in the realspace called Kurdistan.

(Mojab 2001, 56)

As stated in Chapter 2, Tufekci (2017) has since written of the nuanced ways in which activists involved in the 2012 Gezi Park and Taksim Square protests in Istanbul used social media technology.

In academic circles, there is often concern about not falling into the trap of 'technodeterminism' – the simplistic and reductive notion that after Twitter and Facebook were created, their mere existence somehow caused revolutions to happen. Causation in this case is not a question that can easily be answered by selecting one of two binary opposites, either the humans or the technology. Activists used these technologies in sundry notable ways: organizing, breaking censorship, publicizing, and coordinating. Older technologies would not have afforded them the same options and would likely have caused their movements to have different trajectories. Technology influences and structures possible outcomes of human action, but it does so in complex ways and never as a single, omnipotent actor. . . .

(119)

Lessons from STS studies about the limits of "technological determinism" (Pacey 1983, 24) are salient, along with recognition that social skills are essential when using technology. The concept of socio-technical systems theory – normally applied to workplaces (Haddad 2002, 41–45) – has some relevance here.

Gender-based obstacles encountered and overcome

In all of the case studies, women described gender-based obstacles they had to surmount in their cultures more generally and/or when engaging in work and technologies considered to be the province of men. Some of the dragon boat builders lived in households and communities with traditional divisions of labor, and career choices for many growing up in the 1950s were narrowly defined, especially given societal expectations to marry and raise children. The Egyptian feminist cyberactivists discussed strident opposition to Internet postings, sexual harassment at protest rallies, and simplistic interpretations of their work by Western media sources.

Female audio engineers doing studio work encountered resistance from male clients, and at certain studios were paid lower wages than their male counterparts. Norma, the Detroit water utility worker, found that young male members of her crew stated that the work was no place for a woman. The female pilots experienced prejudice either from aviation instructors, from customers who did not want to fly with them, and in one case from a higher-ranking co-worker who had less commercial flying experience. Pilots with children often deferred promotion to Captain in order to have more flexible schedules.

In each situation, the women who endured criticism persisted by working hard and excelling in what they did, which served to educate those with whom they worked; by networking with one another for support and in some cases receiving encouragement from spouses and other family members; and by focusing on their passion for the work or cause in which they were engaged. They also made use of support networks outside of their work organizations as with Norma's involvement with tradeswomen in the Detroit area, the audio engineers' networking through women's music festival and feminist circles, and the pilots' connection with organizations like the Ninety-Nines. Wright (2016) discussed the importance of networks for women in male-dominated workplaces, and while the focus of her article was on sexual minorities, the same is true for women of all sexual orientations.

Personal benefits derived from technology mastery

There were a number of extrinsic and intrinsic benefits that women in this study derived from their work with technology. The dragon boat builders were empowered by their mastery of woodworking tools and techniques, and realized newfound support from one another and from the community as they pursued their work. They also gained increased confidence and willingness to share their technical knowledge with other women and apply it in other domains of their lives. The Egyptian women related that their political work was empowering and also stressful, but ICTs and other methods of organizing linked them to like-minded people locally and globally. The feminist audio engineers found a way to integrate their love of music with the technical and artistic aspects of their work, and to serve as role models and mentors to other women.

Norma's work for the Detroit Water Board connected her with other trades-women in the area and led to a life-changing experience as she met and connected in a very personal way with Mongolian residents of Kazakhstan. The commercial airline pilots were fascinated by the technology and enjoyed the challenge of mastering flying skills in a variety of aircraft and under changing atmospheric conditions. That mastery led to increased personal confidence and pride in their recognition that they were and remain a minority in a profession dominated by men. Moreover, in addition to tangible benefits like discounts on air flights for family members and personal travel opportunities, two of them described the pure joy and amazement that comes from what they see from the air.

Implications for women's technology mastery

Toward a new paradigm

The case studies in this volume offer new insights about women's empowered use of non-traditional technology. One is that the motivation to engage with technology was internal, sometimes beginning in childhood. Although some women were encouraged by a role model, teacher, family member, or peers – of either gender – they were self-driven in their determination to master technology use. Once they did so, they often took on the role of mentoring other women who are interested in learning the technology trade. In all cases, the learning and mastery of the technological equipment and tools was a means to another personal goal, and not an end unto itself. Academic programs in technology fields stand to benefit from the recognition that interest in technology for some girls begins early in life, and there is value in assessing and building upon that interest – as early as primary school. STEM mentoring programs offered at postsecondary institutions may miss that cohort. The "STEM Initiative" sponsored by the Baltimore Community Downtown Sailing Center is a wonderful example of a partnership with elementary and middle schools that combines STEM coursework with sailing instruction (Downtown Sailing Center 2018).

Once those girls with an interest in technology are identified, they must be encouraged throughout all levels of schooling, and support networks must extend into professional careers. Given inhospitable work climates for women in technical workplaces (Armstrong, Riemenschneider, and Giddens 2018;

Reilly, Rackley, and Awad 2017) and in society at large, informal and formal support networks of the type described in these case studies are vital to women's persistence and success. Moreover, economic security has been found to be another contextual factor influencing "women's engagement in STEM education and occupations"—ironically in countries with low levels of gender equality (Stoet and Geary 2018, 581). Institutions seeking to attract girls and women to STEM may find it beneficial to identify those who wish to exert economic control over their lives, even if they envision marriage or life partnerships in their futures.

A second insight is that the ways in which women developed technology proficiency were for the most part informal, rather than in school or classroom settings. Learning in group settings like the dragon boat build was a cooperative endeavor, with women helping one another to solve problems and share tools. The audio engineers also shared their knowledge with one another – even when separated geographically.

The skills required to use the varied technologies were socio-technical in nature, with technology in certain instances playing a subordinate role to human skills – sensory and organizational. Two of the audio engineers broadened socio-technical theory even further by describing the added artistic nature of mixing music. In a similar fashion, the dragon boat builders worked to make their creation not only seaworthy, but beautiful, and once the work was finished, they sought to protect her from scratches by foregoing the proverbial breaking of a champagne bottle on the hull at the point of launch.

Most of the women encountered gender-related obstacles along the way, but persisted by demonstrating their seriousness and willingness to work hard, their focus on high-quality performance, and their courage – demonstrated vividly by the Egyptian activists and the woman who worked in the sewers of Detroit. Individual initiative was bolstered by collective support from peers and networks of other women doing similar work. The airline pilots and Egyptian feminists were inspired by those women who had gone before them. In all of the other cases, the women themselves were the pioneers. Of great significance was that the women's passion for their work and/or cause fueled their determination to push through barriers and carry on.

The role of feminism

All of the Egyptian women interviewed integrated their identities as feminists into their social justice work, much like the women in English's (2002, 247) study. Their volunteer work, paid work, and studies reflected a focus on women's rights as well as human rights. Their experiences contrast with the women Allam (2017) studied to determine "why female protestors did not explicitly voice women's rights and gender equality demands in the 2011 Egyptian uprising" (13). As Wahba (2016) has written, gender was very much tied into the Revolution, arguing that "women's participation and activism in the public sphere helped produce alternative narratives that were central to defying patriarch and triggering social transformation in Egypt" (66).

The audio engineers saw their work as contributing to social change as they expanded the reach of women's music performed on stage and in studios. Like the Egyptian feminists, they went beyond acquiring technical skills merely for personal growth. The intent of both groups was broader societal transformation toward gender equality and democracy, much like that envisioned by Madsen and Cook (2010, 146). And while that may not have been the intent of the dragon boat builders, it was certainly an effect as they traded the role of victim for that of powerful survivor. They also benefitted from gains that had been made by the second-wave feminist movement in Newfoundland and nationally (St. John's Women's Centre 2018; Morris et al. 2006 and 2016; Government of Canada 2018) which helped to redefine women's roles in the province, and also from the political savvy of members like Julie Bettney, Anne Marie Anonsen, and Gerry Rogers.

All of the cases illustrated Faulkner's (2001, 90) notion that women "cannot transform gender relations without engaging in technology." Faulkner elaborated on that point by discussing the need to find "non-threatening ways of enabling women to increase their technological competence so that they are less reliant on men's expertise" and referred explicitly to the "'mend your own car' and IT classes" and "women's self-help health groups" that "sought to develop and share alternatives to medical knowledge and practices" (91). By developing technology competencies and helping one another in that quest, the women in this study were cracking gendered boundaries and by their own example implicitly incorporating feminist values in the technologies used and in the organizations/societies in which the technology was embedded.

The case studies also convey that women's mastery of non-traditional technologies – even though motivated by internal factors – occurred in a larger context in which assumptions about appropriate roles for women were challenged by feminist movements. The breaking down of gendered barriers through earlier and contemporary activism on many fronts – including legal and political – created enough social space for individuals and groups to become proficient technology users, regardless of whether they identified overtly as feminists. This is a powerful lesson for the present time, as efforts to discredit the gains of the feminist movement continue in public discourse – particularly in the U.S. – and some women attribute their socio-economic advancement only to individual talent and perseverance. To truly "ungender" technology, women must support one another on the path to empowerment. Moreover, institutional efforts to encourage technological competence would do well to integrate lessons from the fields of women's and gender studies, STS studies, and adult education into technical curricula.

References

AAUW. 2017, July. Quick facts. Science, Technology, Engineering, and Math (STEM). American Association of University Women. www.aauw.org/files/2017/07/QuickFactsSTEM.pdf. Accessed August 10, 2017.

Abbott, Edith. 1910. *Women in industry: A study in American economic history*. New York: D. Appleton and Company. Reprint. New York: Arno Press, 1969.

Abdelaal, Doaa. 2012, December 13. The day after the Referendum. Cited by Kaitlyn Soligan. *MADRE Press*. www.madre.org/press-publications/blog-post/day-after-referendum. Accessed December 20, 2012.

Abdoun, Safaa and Caspani, Maria. 2012, November 8. Egypt drops controversial article from draft constitution. *Trust Law*. http://news.trust.org/item/20121108131000-y6soq. Accessed December 22, 2012.

Abdulla, Rasha A. 2007. *The Internet in the Arab world: Egypt and beyond*. New York: Peter Lang, 175 pp.

Abol-Komsan, Nehad, ed. 2016, February 17. Egyptian women in 2015 parliamentary elections: Fighters on the individual system warriors on the electoral lists. *Egyptian Center for Women's Rights*. http://ecwronline.org/?p=6788. Accessed August 24, 2016.

Adam, Margie. 1973. Would you like to tapdance on the moon? On 1976 vinyl record album Margie Adam – Songwriter produced by Pleiades Records, Berkeley.

Adhopia, Vic. 2008, April 28. Misdiagnosed: Anatomy of Newfoundland's cancer-testing scandal. *CBC News in Depth*. www.cbc.ca/news2/background/cancer/misdiagnosed.html. Accessed November 19, 2008.

Afro. 2018. Pilot Brenda Robinson. *Afro: The Black Media Authority*. www.afro.com/first-black-military-woman-elected-to-aviation-hall-of-fame/pilot-brenda-robinson-588x330/ . Accessed October 20, 2018.

AFSCME. 1976, April. Detroit faces layoffs and CETA problems. *The Public Employee*, 41(3). American Federation of State, County, and Municipal Employees files at the Walter P. Reuther Library, Detroit, Michigan.

Agora Moderator. 2014, February 3. Women's gains in the Egyptian Constitution of 2014. *Agora-parl.org: Portal for Parliamentary Development*. https://agora-parl.org/news/womens-gains-egyptian-constitution-2014. Accessed August 26, 2016.

Ahmed, Leila. 1999, 2012. *A border passage: From Cairo to America – A woman's journey*. New York and London: Penguin Books.

Ahmed, Rehana L., William Thomas, Douglas Yee, and Kathryn H. Schmitz. 2006. Randomized controlled trial of weight training and lymphedema in breast cancer survivors. *Journal of Clinical Oncology*, 24(18): 2765–2772.

Airlines for America. 2018, February 1. US Airlines celebrate Black History Month. *Airlines for America blog*. http://airlines.org/blog/u-s-airlines-celebrate-black-history-month/. Accessed July 16, 2018.

Al-Ali, Nadje S. 2002. *Women's movements in the Middle East: Case studies of Egypt and Turkey*. Geneva: United Nations Research Institute for Social Development. https://eprints.soas.ac.uk/4889/2/UNRISD_Report_final.pdf. Accessed February 15, 2013.

Alive. 1981. Wild women don't get the blues. On vinyl record album Call it Jazz produced by Redwood Records, Oakland, California.

Al-Khamri, Hana. 2018, June 24. Why did Saudi Arabia lift the driving ban on women only now? *Al Jazeera.com*, Opinion/Women's Rights. www.aljazeera.com/indepth/opinion/saudi-arabia-lift-driving-ban-women-180621203632446.html. Accessed December 13, 2018.

Allam, Nermin. 2017. *Women and the Egyptian Revolution: Engagement and activism during the 2011 Arab Uprisings*. Cambridge: Cambridge University Press.

al-Sharif, Manal. 2018, May 23. Why I can't go back to Saudi Arabia to drive for the first time with my son. *The Washington Post*, Global Opinions. www.washingtonpost.com/news/global-opinions/wp/2018/05/23/why-i-cant-go-back-to-saudi-arabia-to-drive-for-the-first-time-with-my-son/?noredirect=on&utm_term=.7d1204011506. Accessed December 13, 2018.

Al-Tawy, Ayat. 2013, December 12. Inside Egypt's draft constitution: Progress on key freedoms. *Ahram Online*. http://english.ahram.org.eg/NewsContent/1/64/88458/Egypt/Politics-/Search.aspx?Text=Egypt. Accessed August 24, 2016.

Altobelli, Emma, Leonardo Rapacchietta, Paoli Matteo Angeletti, Luca Barbarante, Filippo Valerio Profeta, and Roberto Fagnano. 2017, April. Breast Cancer Screening Programmes across the WHO European region: Differences among countries based on national income level. *International Journal of Environmental Research and Public Health*, 14(4): 452. www.ncbi.nlm.nih.gov/pmc/articles/PMC5409652/. Accessed August 29, 2018.

amadorsquare. 2012, March 17. Egypt: Where are the women? *amadorsquare blog*, pp. 1–7. https://amadorsquare.wordpress.com/2012/03/17/egypt-where-are-the-women/. Accessed August 14, 2012.

Amaria, Kainaz. 2011, December 21. The "girl in the blue bra". *National Public Radio*, The Picture Show. www.npr.org/sections/pictureshow/2011/12/21/144098384/the-girl-in-the-blue-bra. Accessed August 20, 2012.

American Airlines. 2015. *Female pilots make history*. www.aa.com/i18n/amrcorp/corporateInformation/facts/femalepilots.jsp8. Accessed September 8, 2015.

American Psychological Association (APA). 2014, February. Making air travel safer through crew resource management. *Psychology in Action*. www.apa.org/action/resources/research-in-action/crew.aspx.

Anderson, David M. and Carol J. Haddad. 2005. Gender, voice and learning in online course environments. *Journal of Asynchronous Learning Networks*, 9(1): 3–14.

Anderson, Captain Nina E. 2009. *Flying above the glass ceiling. Inspirational stories of success from the first women pilots to fly airline and corporate aircraft*, edited by William Pratt. Garden City Park, NY: Square One Publishers.

Andres, Norma Jean Lim. 2013, October 17. Recorded interview by Carol Haddad. Detroit, Michigan.

Arab Republic of Egypt. 2015, July. *Measuring the digital society in Egypt: Internet at a glance statistical profile*. Ministry of Communications and Information Technology. www.mcit.gov.eg/Upcont/Documents/Publications_1272015000_Measuring_the_Digital_Society_in_Egypt_12_.pdf

Arab Women Organization. 2016. *Arab Women Organization website, structure: The Supreme Council*. http://english.arabwomenorg.org/ Accessed July 20, 2016.

Arizpe, Lourdes. 1999. Freedom to create: Women's agenda for cyberspace. In *Women @ Internet: Creating new cultures in cyberspace*, edited by Wendy Harcourt, xii–xvi. London & New York: Zed Books.

Armstrong, Deborah J., Cynthia K. Riemenschneider, and Laurie G. Giddens. 2018. The advancement and persistence of women in the information technology profession: An extension of ahuja's gendered theory of IT career stages. *Information Systems Journal*, 28(6): 1082–1124.

Armstrong, Toni, Jr. 1989, January. National women's music festival: Showcase 1988. *Hot Wire*, 5(1): 26.

Associated Press in Beirut. 2016, June 28. Journalist critical of Egyptian government deported from Cairo. *The Guardian*. www.theguardian.com/world/2016/jun/28/lilia ne-daoud-journalist-egypt-deported-cairo-lebanon. Accessed August 24, 2016.

Avalon Dragon Boating. 2008, 2018. *History website*. www.avalondragonboating.com/a valon-dragon-boating-history.html. Accessed July 24, 2008 and on July 11, 2018.

Aviation Week Intelligence Network. 2013, July 12. Editorial: How to end automation dependency. *Aviation Week and Space Technology*. http://aviationweek.com/comm ercial-aviation/editorial-how-end-automation-dependency. Accessed August 14, 2014.

Azzi, Georges. 2011, December 21. *History of the LGBT movement in Lebanon. A Blog by Georges Azzi*. http://gazzi.wordpress.com/2011/12/21/history-of-the-lgbt-movem ent-in-lebanon-3/ Accessed October 28, 2012.

Badran, Margot. 1988. The feminist vision in the writings of three turn-of-the-century Egyptian women. *Bulletin of the British Society for Middle Eastern Studies*, 15(1/2): 11–20. Accessed November 1, 2012.

Badran, Margot. 1992. Feminist politics in early twentieth century Egypt. In *Problems of the Modern Middle East in Historical Perspective*, edited by John P. Spagnolo, 27–48. Reading, UK: Published for the Middle East Centre, St. Antony's College Oxford by Ithaca Press.

Badran, Margot. 1995. *Feminists, Islam, and nation: Gender and the making of modern Egypt*. Princeton, NJ: Princeton University Press, 352 pp.

Baker, Elizabeth Faulkner. 1964. *Technology and women's work*. New York: Columbia University Press.

Baker, Mona, ed. 2015. *Translating dissent: Voices from and with the Egyptian Revolution*. New York and London: Routledge.

Bandura, Albert. 1986. *Social foundations of thought and action: A social cognitive theory*. Englewood Cliffs, NJ: Prentice-Hall.

Banerjee, Sarbani and Amitra Hodge. 2007. Internet usage: A within race analysis. *Race, Gender & Class*, 14(3/4): 228–235, 237–238, 242–246. Accessed April 5, 2010.

Bangalore Bureau. 2009, February 6. We'll not spare dating couples on Valentine's Day: Muthalik. *The Hindu*. www.thehindu.com/todays-paper/article348725.ece. Accessed October 20, 2012.

Barshi, Immanuel. 2015. From Healy's training principles to training specifications: The case of the comprehensive LOFT. *The American Journal of Psychology*, 128 (2): 219–227.

Baxandall, Rosalyn, Linda Gordon, and Susan Reverby, eds. 1976. *America's working women: A documentary history—1600 to the present*. New York: Vintage Books.

BBC. 2011a, January 31. Old technology finds role in Egyptian protests. *BBC News*. www.bbc.co.uk/news/technology-12322948. Accessed October 31, 2012.

BBC. 2011b, February 2. Internet comes back online. *BBC News*. www.bbc.co.uk/news/technology-12346929. Accessed July 1, 2012.

Beiser, Elana. 2017, December 13. Record number of journalists jailed as Turkey, China, Egypt pay scant price for repression. *Committee to Protect Journalists Report*. https://cpj.org/reports/2017/12/journalists-prison-jail-record-number-turkey-china-egypt.php.

Bix, Amy Sue. 2005, September. Bessie Coleman: Race and gender realities behind aviation dreams. In *Realizing the dream of flight: Biographical essays in honor of the centennial flight*, edited by Virginia P. Dawson and Mark D. Bowles. Washington, D. C.: National Aeronautics and Space Administration, NASA History Division, Office of External Relations, 310 pages. https://ntrs.nasa.gov/archive/nasa/casi.ntrs.nasa.gov/20050229888.pdf

Bix, Amy Sue. 2013. *Girls coming to tech! A history of American engineering education for women*. Cambridge, MA: MIT Press.

Black Past. 2017. Brown, Jill E., online encyclopedia entry contributed by Matt Van Houten. *Black Past.org: An online reference guide to African American history*. https://blackpast.org/aah/brown-jill-e-1950. Accessed July 16, 2018.

Boserup, Esther. 1970. *Women's role in economic development*. New York: St. Martin's Press.

Boston Women's Health Book Collective. 1971. *Our bodies, ourselves*. Somerville, MA: New England Free Press.

Brady, Tim. 2000. *The American aviation experience: A history*. Carbondale, IL: Southern Illinois University Press.

Brödner, Peter. 1982. Human work design for man-machine systems: A challenge to engineers and labour scientists. In *Proceedings IFAC/IFIP/IFORS/IEA conference: Analysis, design and evaluation of man-machine systems*, Baden-Baden, Federal Republic of Germany, 179–185. New York: Pergamon Press.

Brown, Jane. 2008, September 17, 9:31 pm. Personal electronic mail message to Carol Haddad.

Bruce, Margaret, Gill Kirkup, and Chris Thomas. 1984, September. *Teaching technology assessment to women*. Milton Keynes: The Open University.

Buhr, Sarah. 2016, Janurary 28. Inside Parlio: Egyptian activist Wael Ghonim's new platform for social change. *Tech Crunch*. https://techcrunch.com/2016/01/28/inside-parlio-egyptian-activist-wael-ghonims-new-platform-for-social-change/ Accessed August 23, 2016.

Bukowski, Diane. 2012, May 21. Detroit founded health department in 1825; it previously ran 3 hospitals including Detroit General, 5 clinics, physician home visit services. *Voice of Detroit: The city's independent newspaper, unbossed and unbought*. http://voiceofdetroit.net/2012/05/21/detroit-founded-health-dept-in-1825-it-previously-ran-3-hospitals-including-detroit-general-5-clinics-physician-home-visit-services/. Accessed March 25, 2017.

Bush, Corlann Gee. 1983. Women and the assessment of technology: to think, to be; to unthink, to free. In *Machina ex dea: Feminist perspectives on technology*, edited by Joan Rothschild, 151–170. New York: Pergamon Press.

Canadian Cancer Society. 2018. *Breast cancer statistics*. www.cancer.ca/en/cancer-information/cancer-type/breast/statistics/?region=on. Accessed August 29, 2018.

Casner, Stephen M., Richard W. Geven, and Kent T. Williams. 2013. The effectiveness of airline pilot training for abnormal events. *Human Factors: The Journal of Human Factors and Ergonomics Society*, 55(3): 477–485.

Casner, Stephen M., Richard W. Geven, Matthias P. Recker, and Jonathan W. Schooler. 2014. The retention of manual flying skills in the automated cockpit. *Human Factors: The Journal of Human Factors and Ergonomics Society,* 56(8): 1506–1516.

Catalyst. 2018, January 3. Women in Science, Technology, Engineering and Mathematics (STEM). www.catalyst.org/knowledge/women-science-technology-engineering-and-ma thematics-stem. Accessed December 15, 2018.

CBC News. 2009, September 28. Outspoken N.L. breast cancer patient dies. *CBC News online.* www.cbc.ca/news/canada/newfoundland-labrador/outspoken-n-l-brea st-cancer-patient-dies-1.843622 Accessed September 15, 2010.

Ceniza, Susan Claire. 1985, Winter. Sisters of invention: Philippines. *Connexions: An International Women's Quarterly,* (15): 24.

Chaney, Elsa M. and Marianne Schmink. 1976. Women and modernization: Access to tools. In *Sex and class in Latin America,* edited by June Nash and Helen Icken Safa. New York: Praeger Publishers.

Charette, Robert N. 2009, December 15. Automated to death: As software pilots more of our vehicles, humans can pay the ultimate price. *IEEE Spectrum.* https://spectrum.ieee. org/computing/software/automated-to-death

Chatterjee, Rhitu. 2018, February 21. A new survey finds 81 percent of women have experienced sexual harassment. *The Two-Way: Breaking News from NPR.* U.S.: National Public Radio. www.npr.org/sections/thetwo-way/2018/02/21/587671849/a -new-survey-finds-eighty-percent-of-women-have-experienced-sexual-harassment. Accessed December 13, 2018.

Cheema, Birinder Singh and Catherine A. Gaul. 2006. Full-body exercise training improves fitness and quality of life in survivors of breast care. *Journal of Strength and Conditioning Research,* 20(1): 14–21.

Chen, Peiying. 2012, October. Empowering identity reconstruction of indigenous college students through transformative learning. *Educational Review,* 64(2): 161–180.

Chicago Women in Trades. 2017. *Chicago Women in Trades website: About us.* http:// chicagowomenintrades2.org/about-us-2/agency-overview/. Accessed March 22, 2017.

Christian, Meg. 1974. Ode to a gym teacher. On 1975 vinyl record album I Know You Know produced by Olivia Records, Los Angeles.

Chronister, Krista M. and Ellen Hawley McWhirter. 2003, Fall. Applying social cognitive career theory to the empowerment of battered women. *Journal of Counseling and Development,* 81: 418–425.

Clark, M. Carolyn and Arthur L. Wilson. 1991, June 1. Context and rationality in Mezirow's theory of transformational learning. *Adult Education Quarterly,* 41(2): 75–91.

Clemmens, Ginni. 1976. Solid Ground. On 1976 vinyl record album I'm Lookin' for Some Long Time Friends produced by Open Door Records.

CLUW. 2019. *Coalition of Labor Union Women website: About CLUW.* www.cluw.org/ index.cfm?zone=/unionactive/view_page.cfm&page=About20CLUW. Accessed January 9, 2019.

Cochrane, Dorothy. 1997. Introduction. In *Women and flight: Portraits of contemporary women pilots,* Carolyn Russo, 10–15. The National Air and Space Museum, Smithsonian Institution in association with Bulfinch Press. Boston: Little, Brown, and Company.

Committee to Protect Journalists. 2018, April 26. *Egypt arrests 3 journalists in 24 hours.* https://cpj.org/2018/04/egypt-arrests-3-local-journalists-in-24-hours.php, Accessed September 24, 2018.

Committee to Protect Journalists. 2015, June 25. *Egypt's imprisonment of journalists is at an all-time high.* https://cpj.org/reports/2015/06/egypt-imprisonment-of-journalists-is-at-an-all-time-high.php. Accessed September 2, 2016.

Committee to Protect Journalists. 2016, April 26. *IPJ Condemns Iran's Jailing of Journalists.* https://cpj.org/2016/04/cpj-condemns-irans-jailing-of-journalists.php. Accessed July 11, 2016.

Constitute Project. 2016, August 19. *Egypt's Constitution of 2014.* www.constituteproject.org/constitution/Egypt_2014.pdf. Accessed August 24, 2016.

Corbett, J. Martin. 1990. Human centred advanced manufacturing systems: From rhetoric to reality. *International Journal of Industrial Ergonomics*, 5: 83–90.

Cullum, Brannon. 2012. The Pink Chaddi Campaign. *Movements.org.* www.movements.org/case-study/entry/the-pink-chaddi-campaign/. Accessed October 5, 2012.

Dalsh, Amr Abdallah. 2012, November 8. Egypt's Constitution drops gender equality article. *A-Monitor.com.* www.al-monitor.com/pulse/politics/2012/10/egypt-constitution-draft-drops-gender-equality-article.html. Accessed December 22, 2012.

Davis, Angela Y. 1998. *Blues legacies and black feminism: Gertrude "Ma" Rainey, Bessie Smith, and Billie Holiday.* New York: Pantheon Books.

Davis, Susan Schaefer. 2007, Winter. Empowering women weavers? The Internet in rural Morocco. *Information Technologies and International Development*, 4(2): 17–23.

de Madres, Manos. 2012, 2016. *Manos de Madres Website: About us.* www.manosmadres.org/AboutUs.asp. Accessed October 30, 2012 and again on June 15, 2016.

Dennehy, Tara C. and Nilanjana Dasgupta. 2017. Female peer mentors early in college increase women's positive academic experiences and retention in engineering. *Proceedings of the National Academy of Sciences.* www.pnas.org/content/114/23/5964.full. Accessed July 17, 2017.

De Roulf, Patty. 1976. *A Woman and Her Car: an easy, illustrated guide to understanding, maintaining, and repairing your car. Dell Purse Book.* New York: Dell Publishing Company.

de Silva-de Alwis, Rangita. 2011. Introduction. *Women Leading Change II: Rabat Roundtable Papers.* Wellesley: Wellesley Centers for Women. www.wcwonline.org/component/option,com_virtuemart/Itemid,175/category_id,480/flypage,flypage.tpl/page,shop.product_details/product_id,1717/. Accessed July 1, 2012.

Dhawan, Himanshi. 2009, February 14. 'Pink chaddi'. campaign a hit, draws over 34,000 members. *The Times of India.* http://articles.timesofindia.indiatimes.com/2009-02-14/india/28037274_1_pink-chaddi-pink-chaddi-campaign-sri-ram-sene. Accessed October 5, 2012.

Direnberger, Lucia. 2011, March. De la rue à Internet : espaces de contestation féminins et feminists à Téhéran [From the street to the internet: feminine and feminist contestation in Teheran] translator: Claire Hancock, *Justice Spatiale | Spatial Justice*, n° 03. www.jssj.org/wp-content/uploads/2012/12/JSSJ3-12en.pdf

Dobkin, Alix. 1973. Talking Lesbian. On 1975 vinyl record album Lavender Jane Loves Women produced by Alix Dobkin, Kay Gardner, and Marilyn Ries, New York.

Douglas, Deborah G. 2004. *American women and flight since 1940.* Lexington: The University Press of Kentucky.

Downtown Sailing Center. 2018. *Downtown Sailing Center's STEM Initiative.* Downtown Sailing Center at the Baltimore Museum of Industry. https://downtownsailing.org/STEM Accessed December 15, 2018.

Drydyk, Jay. 2013. Empowerment, agency, and power. *Journal of Global Ethics*, 9(3): 249–262. DOI: doi:10.1080/17449626.2013.818374. Accessed July 6, 2017.

Duley, Margaret. 1949. *The caribou hut: The story of a Newfoundland hostel*. Toronto: The Ryerson Press.

Duley, Margot I. 1993. *Where once our mothers stood we stand: Women's suffrage in Newfoundland, 1890–1925*. Charlottetown, PEI: Gynergy.

Ebo, Bosah. 1998. Internet or outernet? In *Cyberghetto or cyberutopia? Race, class, and gender on the Internet*, edited by B. Ebo, 1–12. Westport, CT: Praeger.

Eccles [Parsons], Jacquelynne, T. F. Adler, R. Futterman, S. B. Goff, C. M. Kaczala, J. L. Meece, and C. Midgley. 1983. Expectations, values and academic behaviors. In *Perspective on achievement and achievement motivation*, edited by J. T. Spence, 75–146. San Francisco, CA: W. H. Freeman.

Eccles, Jacquelynne S. 1987. Gender roles and women's achievement-related decisions. *Psychology of Women Quarterly*, 11(2): 135–171.

Eccles, Jacquelynne S. 2011. Gendered educational and occupational choices: Applying the eccles et al. model of achievement-related choices. *International Journal of Behavioral Development*, 35(3): 195–201.

Egypt Independent. 2010, June 16. Two witnesses affirm Alex victim beaten by police. www.egyptindependent.com/two-witnesses-affirm-alex-victim-beaten-police/. Accessed December 20, 2012.

Egypt Ministry of Communications and Information Technology. 2009, March. Egypt's Free Internet Initiative provides all Egyptians with free access to the Internet. *Information and Communications Technology Bulletin*, http://modernegypt.info/one-hun dred-facts-about-egypt/fact/32/. Accessed October 30, 2012.

Egyptian Center for Women's Rights. 2014, March 26. Out of a Gunpowder barrel: Egyptian Women's Status Report in 2013. Edited by Nihad Abol-Qomsan. https:// ecwronline.org/?p=4578. Accessed October 20, 2017.

Ehn, Pelle. 1989a. *Work-oriented design of computer artifacts*. Stockholm: Arbetslivs-centrum. Reprinted in 1990 by L. Erlbaum Associates.

Ehn, Pelle. 1989b. The art and science of designing computer artifacts. *Scandinavian Journal of Information Systems*, 1: 21–42.

El Deeb, Sarah. 2012a, June 7. In 'New Egypt' mobs sexually assault women with impunity. *NBC News*. www.nbcnews.com/id/47717050/ns/world_news-mideast_n_a frica/t/new-egypt-mobs-sexually-assault-women-impunity/. Accessed June 11, 2012.

El Deeb, Sarah. 2012b, December 16. Mistrust runs deep over Egypt referendum. *Yahoo News*. www.yahoo.com/news/mistrust-runs-deep-over-egypt-referendum-221724782. html. Accessed December 16, 2012.

El-Ghobashy, Mona. 2005. The metamorphosis of the Egyptian Muslim Brothers. *International Journal of Middle East Studies*, 37(3): 373–395.

El-Naggar, Omar. 2011, December 28. *Egyptian women are a red line that cannot be crossed, YouTube video*. www.youtube.com/watch?v=2E-FD98seog. Accessed December 22, 2012.

El-Nawawy, M. A. 2000. Profiling Internet users in Egypt: Understanding the primary deterrent against their growth in number . *INET 2000 Proceedings*. www.isoc.org/ inet2000/cdproceedings/8d/8d_3.htm. Accessed October 29, 2012.

El Tahawy, Randa. 2013, September 29. Egypt rights groups demand quotas for women in parliament. *Al Arabiya English*. http://english.alarabiya.net/en/persp ective/features/2013/09/29/Calls-rise-for-quota-of-Egyptian-women-parliament.html. Accessed November 10, 2013.

Elliott, Fara. 1995, Aug/Sept. Forget the boys and online porn. It's time for...Cyberfe-minism! *Off Our Backs*, 25(8): 6. Accessed September 22, 2012.

Ellul, Jacques. 1964. *The technological society*. New York: Vintage Books.

England, Andrew and Heba Saleh. 2011, February 9. Google worker is Egypt's Facebook hero. *Financial Times*.

English, Leona M. 2002. Learning how they learn: International adult educators in the global sphere. *Association for Studies in International Education*, 6: 230–248.

Erenrich, Susan J. and Jon P. Wergin, eds. 2017. *Grassroots leadership and the arts for social change*. Bingley, UK: Emerald Publishing Limited.

Euromed Rights. 2018, March 20. Egypt: Closure of the office Nazra for Feminist Studies. News. *Euromed Rights Egypt Press Release: Shrinking Space for Civil Society*. https://euromedrights.org/publication/egypt-closure-office-nazra-feminist-studies/. Accessed September 24, 2018.

Everett, Anna. 2004. On cyberfeminism and cyberwomanism: High-tech mediations of feminism's discontents. *Signs: Journal of Women in Culture and Society*, 30(1): 1278–1286. Accessed September 22, 2012.

Ezzat, Dina. 2012. Breaking the silence: Women protesters in Tahrir Square are calling for an end to political and physical harassment. *Al-Ahram Weekly*, Egypt. http://weekly.ahram.org.eg/Archive/2012/1102/eg13.htm. Accessed June 19, 2012.

Farid, Sonia. 2013, October 7. Tracking Egypt's Islamic identity in the constitution. *Al-Arabiya English*. http://english.alarabiya.net/en/perspective/analysis/2013/10/07/Tracking-Islamic-identity-in-Egypt-s-constitution.html. Accessed August 24, 2016.

Faulkner, Wendy. 2001. The technology question in feminism: A view from feminist technology studies. *Women's Studies International Forum*, 24(1): 79–95.

Federal Aviation Administration. 2012, April 5. The history of CRM. *FAA TV*. www.faa.gov/tv/?mediaId=447

Federal Aviation Administration. 2017. *U.S. Civil Airmen Statistics, 2017*. www.faa.gov/data_research/aviation_data_statistics/civil_airmen_statistics/. Accessed January 8, 2019.

Feldman, Maxine. 1969. Angry Atthis. First released in 1972 as 45 rpm vinyl record, Harrison & Tyler Productions. www.queermusicheritage.us/apr2002.html. Accessed August 7, 2010.

Fox, Mary Frank, Deborah G. Johnson, and Sue V. Rosser, eds. 2006. *Women, gender, and technology*. Preface, vii. Urbana: University of Illinois Press.

Frank, Miriam. 2001. Hard hats & homophobia: Lesbians in the building trades. *New Labor Forum*, (8): 25. http://ezproxy.emich.edu/login?url=http://search.proquest.com/docview/237233957?accountid=10650. Accessed April 1, 2017.

Freire, Paulo. 1971. *Pedagogy of the oppressed*. New York: Continuum Publishing Company.

Freydberg, Elizabeth Amelia Hadley. 1994. *Bessie Coleman: The brownskin lady bird*. New York & London: Garland Publishing, Inc.

Gaar, Gillian G. 1992. *She's a rebel: The history of women in rock & roll*. Seattle: Seal Press.

Galanis, Nikolas, Enric Mayol, Marc Alier, Francisco José García-Peñalvo. 2016, February. Supporting, evaluating and validating informal learning. A social approach. *Computers in Human Behavior*, 55(A): 596–603.

Ganson, Barbara. 2014, Fall. Taking commemorative flight. *Phi Kappa Phi Forum*. www.phikappaphiforum-digital.org/phikappaphiforum/fall_2014?pg=5#pg5

Garlington, Lee. 1977, August 31. 4th national women's music festival: the sounds, set-up, and final days of Be Be K'Roche. *Off Our Backs*, 7(6): 19.

George, Fred. 2014, February 1. Automation dependency: Are pilots becoming "Children of the Magenta?" *Business and Commercial Aviation*, 110(2): 20.

Gheytanchi, Elham. 2015. Gender roles in the social media world of Iranian women. In *Social Media in Iran: Politics and Society after 2009*, edited by David M. Faris and Babak Rahimi, 41–55. Albany, NY: State University of New York Press.

Giglio, Mike. 2011, February 13. How Wael Ghonim sparked a Egypt's uprising. *Newsweek*. www.newsweek.com/how-wael-ghonim-sparked-egypts-uprising-68727. Accessed December 20, 2012.

Gjersoe, Nathalia. 2018, March 8. Bridging the gender gap: why do so few girls study Stem subjects? *The Guardian*. www.theguardian.com/science/head-quarters/2018/mar/08/bridging-the-gender-gap-why-do-so-few-girls-study-stem-subjects. Accessed December 15, 2018.

Gladstone, Rick and Artin Afkhami. 2012, January 25. Pattern of intimidation is seen in arrests of Iranian journalists and bloggers. *The New York Times*. www.nytimes.com/2012/01/26/world/middleeast/iran-steps-up-arrests-of-journalists-and-bloggers.html?ref=middleeast

Global Fund for Women. 2016, March. Stop the campaign of repression against Mozn Hassan and all Human Rights Defenders in Egypt. *Global Fund for Women News*. www.globalfundforwomen.org/mozn-hassan-stop-repression-human-rights-defenders-egypt/#.W6kjmHtKh0x. Accessed September 24, 2018.

Goldstein, Joshua and Juliana Rotich. 2008, September. *Digitally Networked Technology in Kenya's 2007–2008 Post-Election Crisis*. Publication No. 2008–2009. Cambridge, MA: The Berkman Center for Internet and Society at Harvard University. https://cyber.harvard.edu/sites/cyber.harvard.edu/files/Goldstein&Rotich_Digitally_Networked_Technology_Kenyas_Crisis.pdf.pdf . Accessed October 11, 2012.

Government of Canada. 2018, November 22. *Guide to the Canadian Charter of Rights and Freedoms*. www.canada.ca/en/canadian-heritage/services/how-rights-protected/guide-canadian-charter-rights-freedoms.html. Accessed December 28, 2018.

Gray, Jessica. 2012, January 30. Egypt's Feminist Union undergoing reincarnation. *Women's e-news*. https://womensenews.org/2012/01/egypts-feminist-union-undergoing-reincarnation/#.UJl81MXqmSo. Accessed February 21, 2012.

Green, Eileen, Jenny Owen, and Den Pain, eds. 1993. *Gendered by design? Information technology and office systems*. London: Taylor and Francis.

Green, Jeff, ed. 2008. Winter. *The Communicator: Memorial University's Employee Newsletter*, 22(4): 1–12. www.mun.ca/marcomm/work/Communicator_Winter_2008.pdf

Gulhane, Joel. 2012, December 15. UN experts concerned over women's rights in proposed constitution. *Daily News Egypt*. https://dailynewsegypt.com/2012/12/15/un-experts-concerned-over-womens-rights-in-proposed-constitution/. Accessed January 2, 2012.

Gutierrez, Lorraine M. 1995. Understanding the empowerment process: Does consciousness make a difference? *Social Work Research*, 19: 229–237.

Habib, Heba. 2016, May 31. Egypt detains leadership of journalists union on charges of harboring fugitives. *The Washington Post*. www.washingtonpost.com/world/egypt-detains-leadership-of-journalists-union-on-charges-of-harboring-fugitives/2016/05/31/71f04ee8-2716-11e6-ae4a-3cdd5fe74204_story.html?noredirect=on&utm_term=.cadd45dd491d. Accessed August 24, 2016.

Haddad, Carol J. 1987. Technology, industrialization, and the economic status of women. In *Women, work and technology: Transformations*, edited by Barbara Drygulski Wright, et. al., 33–57. Ann Arbor: The University of Michigan Press.

Haddad, Carol J. 1989. *Technology and skill: Educational considerations in the implementation and use of advanced manufacturing technology* (Doctoral dissertation). Ann Arbor: The University of Michigan.

Haddad, Carol J. 2002. *Managing technological change: A strategic partnership approach.* Thousand Oaks, CA: Sage Publications.

Hamilton, Mary. 2006, October. Just do it: Literacies, everyday learning and the irrelevance of pedagogy. *Studies in the Education of Adults,* 38(2): 125–140.

Hamlin, Jesse. 2007, February 9. Grammy winner's sound advice / Daughter of Spike Jones has mixed, mastered and miked everyone from Santana to Kronos Quartet. Listen up. *SF Gate.* www.sfgate.com/entertainment/article/Grammy-winner-s-sound-a dvice-Daughter-of-Spike-2650332.php.

Hannon, John. 2013, April. Incommensurate practices: Sociomaterial entanglements of learning technology implementation. *Journal of Computer Assisted Learning,* 29(2): 109–206.

Haraway, Donna J. 1991. *Simians, cyborgs, and women: The reinvention of nature.* New York: Routledge.

Hardesty, Von. 2008. *Black wings: courageous stories of African Americans in aviation and space history.* New York: HarperCollins Publishers, in association with the National Air and Space Museum.

Harris, Susan R. 2001, Jan. 23. Clinical practice guidelines for the care and treatment of breast cancer: 11. Lymphedema. *Canadian Medical Association Journal,* 164(2): 191–199.

Harris, Susan R. 2008. Exercise and breast cancer: Why the Dragon Boat experience is important for women living with breast cancer. www.bustingwithenergy.com/history. html. Accessed September 10, 2016.

Harris, Susan R. 2012. "We're all in the same boat": A review of the benefits of dragon boat racing for women living with breast cancer. *Evidence-Based Complementary and Alternative Medicine,* 2012(167651): 1–6.

Harty, Jack. 2014, August. Best of Airways – Texas International Airlines. *Airways Magazine.* https://airwaysmag.com/avgeek/best-of-airways-texas-international-airlines/ . Accessed July 16, 2018.

Hassanin, Leila. 2007. *Global Information Society Watch: Focus on Participation, Egypt.* Association for Progressive Communications and Third World Institute. Global Information Society Watch. www.giswatch.org/sites/default/files/gisw_egypt.pdf Accessed October 29, 2012.

Hassi, Marja-Liisa and Sandra L. Laursen. 2015, October. Transformative learning: Personal empowerment in learning mathematics. *Journal of Transformative Education,* 13(4): 316–340.

Hatem, Mervat F. 1992, May. Economic and political Liberation in Egypt and the demise of state feminism. *International Journal of Middle East,* 24(2): 231–251. Accessed November 3, 2012.

Helem. 2017. *Official Facebook Page for Helem.* www.facebook.com/Official-Page-for-Helem-Lebanon-133916233311662/. Accessed July 24, 2017.

Herrera, Linda. 2014. *Revolution in the age of social media: The Egyptian popular insurrection and the Internet.* London and New York: Verso Books.

Hess, Edward D. and Noah Arlow. 2014. *Learn or die: Using science to build a leading-edge learning organization.* New York: Columbia Business School Publishing.

Hicks, Mitti. 2018, December 14. Meet Beth Powell, an American Airlines pilot encouraging others to reach new heights. *Travel Noire.* https://travelnoire.com/beth-p owell-american-airlines-pilot/. Accessed January 8, 2019.

Historica Canada. 2013, October. Newfoundland and Labrador and Confederation. *The Canadian Encyclopedia.* www.thecanadianencyclopedia.ca/en/article/newfoundland-a nd-labrador-and-confederation/. Accessed March 19, 2017.

Hodges, Mark. 1998, August. Averting aviation disasters. *Computer Graphics World*, 21.

Holmes, Michelle D., Wendy Y. Chen, Diane Feskanich, Candyce H. Kroenke, and Graham A. Colditz. 2005, May 25. Physical activity and survival after breast cancer diagnosis. *JAMA, The Journal of the American Medical Association*, 293(20): 2479–2487.

Hoover, Eric. 2000, October 13. Fading places. The closing of a women's haven leaves the lesbian community short on institutions. *Washington City Paper*. www.washingtoncitypaper.com/news/article/13020943/fading-places. Accessed May 14, 2017.

Hornblower, Margot. 2001, June 24. The still unfriendly skies. *Time*. http://content.time.com/time/printout/0,8816,134616,00.html. Accessed July 16, 2018.

Human Rights Watch. 2016, June 28. *Egypt: Travel ban on women's rights leader*. www.hrw.org/news/2016/06/28/egypt-travel-ban-womens-rights-leader. Accessed August 24, 2016.

IBCPC. 2014. *IBCPC Participatory Dragon Boat Festival, Sarasota, Florida, USA*. www.sarasotabcs2014festival.org/. Accessed September 5, 2015.

International Dragon Boat Federation. 2008. The dragon boat: History and Culture. *IDBF webpage*. www.idbf.org/documents/DB_History_Culture.pdf. Accessed October 26, 2008.

International Foundation for Electoral Systems. 2011, November 1. *Analysis of the 2011 Parliamentary Electoral System*. Washington, D.C.: Middle East and North Africa International Foundation for Electoral Systems. www.ifes.org/publications/analysis-egypts-2011-parliamentary-electoral-system. http://aceproject.org/ero-en/regions/africa/MZ/egypt-analysis-of-the-2011-parliamentary-electoral. Accessed November 6, 2012.

International Society of Women Airline Pilots. 2018. *Frequently Asked Questions*. https://s3.amazonaws.com/ClubExpressClubFiles/658242/documents/Worldwide_List_of_Airline_Numbers_and_Women_Pilots_December_17_2018_331878125.pdf?AWSAccessKeyId=AKIAIB6I23VLJX7E4J7Q&Expires=1546968236&response-content-disposition=inline%3B%20filename%3DWorldwide_List_of_Airline_Numbers_and_Women_Pilots_December_17_2018.pdf&Signature=T7jr0K8hkYoug%2Bm4ppvMzBeYe64%3D. Accessed January 8, 2019.

International Telecommunication Union (ITU). 2018, December 7. Press Release: ITU releases 2018 global and regional ICT estimates. *ITU*, Geneva, Switzerland. www.itu.int/en/mediacentre/Pages/2018-PR40.aspx. Accessed December 11, 2018.

Johnson, Deborah G. 2006. Introduction. In *Women, gender, and technology*, edited by Mary Frank Fox, Deborah G. Johnson, and Sue V. Rosser, 1–11. Urbana: University of Illinois Press.

Jones, Steve. 2002. *Cyberfeminism. Encyclopedia of new media: An essential reference guide to communication and technology*. Thousand Oaks, CA: Sage Publications.

Kamali Dehghan, Saeed. 2014, July 8. Iranian reporter sentenced to two years in prison and 50 lashes. *The Guardian*. www.theguardian.com/world/2014/jul/08/iran-journalist-prison-lashes-propaganda-government?CMP=twt_gu. Accessed July 22, 2016.

Kamel, Sherif. 1997. The birth of Egypt's information society. *International Journal of the Computer, the Internet, and Management*, 5(3): 7–28.

Kamel, Sherif. 2004. *Evolution of mobile technology in Egypt*. Idea Group Publishing. www.irma-international.org/viewtitle/32466/. Accessed October 29, 2012.

Kamel, Tarek. 1997. *The Internet commercialization in Egypt: Challenges and opportunities*. www.isoc.org/inet97/ans97/tarek.html. Accessed October 29, 2012.

Kane, Karen. 2017. *Karen Kane website*. www.mixmama.com/bio.html, Accessed in 2014 and April 10, 2017.

Karimi, Sedigheh. 2018, March. *The virtual sphere and the women's movement in post-reform Iran* (Doctoral dissertation). Melbourne, Australia. The University of Melbourne, Asia Institute. https://minerva-access.unimelb.edu.au/bitstream/handle/11343/213542/The%20Virtual%20Sphere%20and%20the%20Iranian%20Women%27s%20movment.pdf?sequence=1&isAllowed=y. Accessed December 13, 2018.

Kenyan Pundit. 2012. *Kenyan Pundit Website: About Us.* www.kenyanpundit.com/about/

Kessler-Harris, Alice. 1982. *Out to work: A history of wage-earning women in the United States.* New York: Oxford University Press.

Kessler, Mark. 2018, November 30. Breast cancer and dragon boat racing: The story behind a movement. *WBUR.org.* www.wbur.org/onlyagame/2018/11/30/sandy-smith-mckenzie-harris-frost

Khalid, Osama. 2012, October 14. Saudi Arabia: Women2Drive steps up tone, blames government. *Global Voices.* http://globalvoicesonline.org/2012/10/14/saudi-arabia-women2drive-steps-up-tone-blames-government-policies/. Accessed October 25, 2012.

Kirkpatrick, David D. 2014, May 29. International observers find Egypt's presidential election fell short of standards. *The New York Times.* www.nytimes.com/2014/05/30/world/middleeast/international-observers-find-fault-with-egypt-vote.html. Accessed August 24, 2016.

Kitchenham, Andrew. 2008, April. The evolution of John Mezirow's transformative learning theory. *Journal of Transformative Education,* 6(2): 104–123.

Lane, K., D. Worsley, and D. McKenzie. 2005. Exercise and the lymphatic system: Implications for breast-cancer survivors. *Sports Medicine,* 35(6): 461–471.

LaTour, Jane. 2014. *Talking history. Sisters in the brotherhood: Carpenter Ronnie Sandler.* Oral interview on March 26, 1995. www.talkinghistory.org/sisters/sandler.html. Accessed March 22, 2017.

Lebow, Eileen F. 2002. *Before Amelia: Women pilots in the early days of aviation.* Washington, D.C.: Potomac Books, Inc.

Le Clus, Megan. 2011, July. Informal learning in the workplace: A review of the literature. *Australian Journal of Adult Learning,* 51(2): 355–373. Accessed December 12, 2018.

Lee, Courtland C. 2007, April 1. Empowerment theory for the professional school counselor: A manifesto for what really matters. *Professional School Counseling,* 1096–2409. Accessed March 27, 2008.

Lems, Kristin. 1978. Mammary glands. On 1978 vinyl record album Oh Mama! produced by Carolsdatter Productions, Urbana, IL.

Lems, Kristin. 1982, 2007. Ballad of the E.R.A. http://kristinlems.com/equality_road___double_cd/s/ballad_of_the_era

Leonard, Eileen B. 2003. *Women, technology, and the myth of progress.* Upper Saddle River, N.J.: Prentice Hall.

Lerman, Nina E. 2010, October. Categories of difference, categories of power: Bringing gender and race to the history of technology. *Technology and Culture,* 51(4): 893–918.

Liker, Jeffrey K., Carol J. Haddad, and Jennifer Karlin. 1999. Perspectives on technology and work organization. *Annual Review of Sociology,* 25, 575–596.

Lippitt, Jill. 1981, August 5. Personal communication to the author on behalf of the National Women's Mailing List.

Livingstone, David W. 2000. *Exploring the icebergs of adult learning: Findings of the first Canadian survey of informal learning practices.* NALL Working Paper No. 10.

Lo, Vincent. 2008. *Six-Sixteen Dragon Boats Ltd website.* www.616.ca/. Accessed October 26, 2008.

Lohan, Maria. 2000, December. Constructive tensions in feminist technology studies. *Social Studies of Science*, 30(6): 895–916.

Londono, Ernesto. 2011, February 9. Egyptian man's death became symbol of callous state. *The Washington Post*. www.washingtonpost.com/wp-dyn/content/article/2011/02/08/AR2011020806360.html

Lubar, Steven. 1998. Men, women, production, consumption. In *His and hers: gender, consumption, and technology*, edited by Roger Horowitz and Arwen Mohun. Charlottesville: University Press of Virginia.

M, Nadine. 2010, March 14. Arab queer women and transgenders confronting diverse religious fundamentalisms: The case of Meem in Lebanon. *AWID Women's Rights*. Association for Women's Rights in Development. https://issuu.com/awid/docs/arab_queer_women_and_transgenders_confronting_dive?viewMode=presentation&layout=http://skin.issuu.com/v/light/layout.xml&showFlipBtn=true&e=2350791/8513474. Accessed July 12, 2016.

Macha, Ndesanjo. 2008, January 15. Kenya: Cyberactivism in the aftermath of political violence. *Global Voices*. http://globalvoicesonline.org/2008/01/15/kenya-cyberactivism-in-the-aftermath-of-political-violence/

Madsen, Susan R. and Bradley J. Cook. 2010. Transformative learning: UAE, women, and higher education. *Journal of Global Responsibility*, 1(1): 127–148. Accessed December 10, 2018.

Malik, Nesrine. 2011, June 3. Saudi Arabia's Women2Drive campaign is up against society. *The Guardian*. www.guardian.co.uk/commentisfree/2011/jun/03/saudi-arabia-women2drive-women-driving. Accessed October 25, 2012.

Manuti, Amelia, Serafina Pastore, Anna Fausta Scardigno, Maria Luisa Giancaspro, and Daniele Morciano. 2015, March. Formal and informal learning in the workplace: A research review. *International Journal of Training and Development*, 19(1): 1–17. Accessed December 4, 2018.

Margolis, Jane and Allan Fisher. 2002. *Unlocking the clubhouse: Women in computing*. Cambridge, MA: MIT Press.

Mårtensson, Lena. 1995. The aircraft accident at Gottröra–the experiences of the cockpit crew. *The International Journal of Aviation Psychology*, 5(3): 308–332.

Mårtensson, Lena. 1996. Are operators and pilots in control of complex systems? *Control Engineering Practice*, 7(2): 173–182.

McCullough, Joan. 1980. *First of all: Significant firsts by American women*. New York: Henry Holt and Company.

McGaw, Judith A. 1982. Women and the history of American technology. *Signs*, 7(4): 798–828.

McKenzie, Donald C. and Andrea L. Kalda. 2003. Effect of upper extremity exercise on secondary lymphedema in breast cancer patients: A pilot study. *Journal of Clinical Oncology*, 21(3): 463–466.

McLarney, Ellen. 2016, April 28. Women's equality: Constitutions and revolutions in Egypt. *Project on Middle East Political Science*. https://pomeps.org/2016/04/28/womens-equality-constitutions-and-revolutions-in-egypt/. Accessed August 24, 2016.

McKnight, Jennie. 1987, May 24–31. Volunteers spark Sisterfire festival: Making music happen. *Gay Community News*, 14(43): 7.

Meem. 2010, August 27. *Meem Website: What is Meem?*http://meemgroup.org/news/what-is-meem/. Accessed October 28, 2012; website no longer active.

Meem. 2017. *Meemblog*. https://twitter.com/meemblog. Accessed July 24, 2017.

Megahed, Horeya T. 2014, May 2. Egyptian women and the new constitution. *International Alliance of Women*. https://womenalliance.org/egyptian-women-and-the-new-constitution. Accessed August 24, 2016.

Meyer, Alan. 2015. *Weekend pilots: Technology, masculinity, and private aviation in postwar America*. Baltimore, MD: Johns Hopkins University Press.

Meyer-Resende, Michael. 2014, April. *Egypt: In-depth analysis of the main elements of the new constitution*. European Union, Directorate-General for External Policies of the Union, Directorate B, Policy Department. www.europarl.europa.eu/RegData/etudes/note/join/2014/433846/EXPO-AFET_NT(2014)433846_EN.pdf. Accessed August 24, 2016.

Mezirow, Jack. 1994, December. Understanding transformation theory. *Adult Education Quarterly*, 44(4): 222–232.

Mezirow, Jack. 1997, Summer. Transformative learning: Theory to practice. *New Directions for Adult and Continuing Education*, (74): 5–12.

Mezirow, Jack. 2000. *Learning as transformation: Critical perspectives on a theory in progress*. San Francisco: Jossey-Bass. Cited in Kitchenham, 2008.

Michigan Oral History Database Project. 2004, June 24. *Eva Caradonna Oral History*. Wayne State University, Walter P. Reuther Library. http://xserve2.reuther.wayne.edu/SPT–FullRecord.php?ResourceId=1864. Accessed March 31, 2017.

Michigan Women's Hall of Fame. 1998. *Hilda Patricia Curran. Michigan Women's Historical Center and Hall of Fame Contemporary Inductee*. Lansing, Michigan. www.michiganwomenshalloffame.org/Images/Curran,%20Hilda%20Patricia.pdf. Accessed March 22, 2017.

Miedema, Baukje, Ryan Hamilton, Sue Tatemichi, Roanne Thomas-MacLean, Anna Towers, Thomas F. Hack, Andrea Tilley, and Winkle Kwan. 2008. Predicting recreational difficulties and decreased leisure activities in women 6–12 months post breast cancer surgery. *Journal of Cancer Survivorship*, 2(4): 262–268.

Million Women Mentors. 2018, October 1. *MWM Newsletter*. www.millionwomenmentors.com/newsletters. Accessed December 15, 2018.

Misquith, Chethan. 2018, March 13. Mangalore pub attack: Sri Ram Sene chief Pramod Muthalik, 24 others acquitted. *Times of India*. https://timesofindia.indiatimes.com/india/mangalore-pub-attack-sri-ram-sene-chief-pramod-muthalik-24-others-acquitted/articleshow/63272444.cms. Accessed December 15, 2018.

Mohie, Mostafa. 2016, December 14. "Punished for doing their jobs": Lawyers defend H. R. workers in court. *Mada*. www.madamasr.com/en/2016/12/14/feature/politics/punished-for-doing-their-jobs-lawyers-defend-hr-workers/. Accessed September 24, 2018.

Mojab, Shahrzad. 2001. The politics of "cyberfeminism" in the middle east: The case of Kurdish women. *Race, Gender & Class*, 8(4): 42–61. Accessed September 22, 2012.

Morgall, Janine M. 1993. *Technology assessment: A feminist perspective*. Philadelphia: Temple University Press.

Morgan, Robin. 2011, Spring. Women of the Arab Spring. *Ms. Magazine*. XXI(2): 20–22.

Morris, Cerise, Anne-marie Pedersen, Calina Ellwand, and Maude-emmanuelle Lambert. 2006 and 2016. *The Canadian Encyclopedia*. www.thecanadianencyclopedia.ca/en/article/royal-commission-on-the-status-of-women-in-canada. Accessed December 28, 2018.

Morris, Bonnie J. 1998. In their own voices: Oral histories of festival artists. *Frontiers*, 19(2): 53–72.

Morris, Hugh. 2018, September 6. The surprising country with more female pilots than any other. *The Telegraph*. www.telegraph.co.uk/travel/news/india-female-pilots/. Accessed January 8, 2019.

Morris, Patricia T., Suzanne Kindervatter, and Amy Woods. 1999. *The gender audit: A process for organizational self-assessment and action planning.* Commission on the Advancement of Women. Washington, D.C.: InterAction.

Morrow, Adam and al-Omrani, Khaled Moussa. 2012. Women fight for rights in Egypt. *IDN-InDepthNews.* https://archive-2011-2016.indepthnews.net/index.php/component/content/article/7-global-issues/1097-women-fight-for-rights-in-new-egypt. Accessed December 22, 2012.

Mostafa, Dalia Said. 2015. Introduction: Egyptian women, revolution, and protest culture. *Journal for Cultural Research*, 19(2): 118–129.

Mostafa, Dalia Said, ed. 2016. *Women, culture, and the January 2011 Egyptian Revolution.* New York and London: Routledge.

Mouawad, Jad. 2014, September 7. Airlines take the bump out of turbulence. *The New York Times.* www.nytimes.com/2014/09/08/technology/airlines-take-the-bump-out-of-turbulence.html

Mount Clemens Downtown Development Authority. 2013. *Mineral Baths.* www.downtownmountclemens.com/historical/mineral-baths. Accessed March 24, 2017.

Mufta Editors. 2012, February 28. Iranian journalist Marzieh Rasouli released on bail. http://muftah.org/iranian-journalist-marzieh-rasouli-released-on-bail/

Murray, Hilda Chaulk. 1979. *More than fifty percent: woman's life in a Newfoundland outport, 1900–1950.* St. John's: Breakwater Books.

Nasawiya. 2017a. *Nasawiya Twitter.* https://twitter.com/nasawiya. Accessed July 24, 2017.

Nasawiya. 2017b. *Nasawiya website: Our causes.* www.nasawiya.org/our-causes/. Accessed July 24, 2017.

National Girls Collaborative Project. 2018, March. *The state of girls and women in STEM.* https://ngcproject.org/sites/default/files/ngcp_the_state_of_girls_and_women_in_stem_2018a.pdf. Accessed July 9, 2018.

National Naval Aviation Museum. 2018. *Discovery Saturday – The First Black Female Naval Aviator.* www.navalaviationmuseum.org/event/discovery-saturday-the-first-black-female-naval-aviator/. Accessed July 16, 2018.

National WASP WWII Museum. n.d. *Women Pilots of World War II.* 36-page booklet. Sweetwater, TX.

Nazra. 2016, March 28. A solidarity statement from the Egyptian Feminist Organizations Coalition with Nazra for Feminist Studies. *Nazra for Feminist Studies website.* http://nazra.org/en/2016/03/solidarity-statement-egyptian-feminist-organizations-coalition-nazra-feminist-studies. Accessed August 24, 2016.

NBC. 2017, March 1. Retired American Airlines Pilot Broke Aviation Barriers. *NBC,* Dallas-Fort Worth, TX. www.nbcdfw.com/news/local/Retired-American-Airlines-Pilot-Broke-Aviation-Barriers-415143083.html. Accessed July 16, 2018.

Near, Holly. 1978. Fight back. On 1978 vinyl record album Imagine my Surprise, produced by Redwood Records, Oakland, California.

Nelson, Cynthia. 1996. *Doria Shafik, Egyptian feminist: A woman apart.* Gainesville: University Press of Florida.

Nilsen, Sigurd R. 1984, May. Recessionary impacts on unemployment. *Monthly Labor Review*, 21–25. www.bls.gov/opub/mlr/1984/05/art4full.pdf. Accessed January 9, 2019.

Ninety-Nines. 2019. *Ninety-Nines: Who we are.* www.ninety-nines.org/who-we-are.htm Accessed January 8, 2019.

Nisha. 2009, April 11. Hackers. *The Pink Chaddi Campaign Blogspot.* http://thepinkchaddicampaign.blogspot.com/ Accessed October 20, 2012.

Nontraditional Employment for Women. 2017. *Non-Traditional Employment for Women website: About NEW.* Accessed March 22, 2017.

Noonan, Ryan. 2017. *Women in STEM: 2017 Update.* U.S. Department of Commerce, Economics and Statistics Administration, Office of the Chief Economist. www.commerce.gov/sites/commerce.gov/files/migrated/reports/women-in-stem-2017-update.pdf. Accessed August 14, 2018.

NWML (National Women's Mailing List). n.d.*National Women's Mailing List brochure.* Jenner, CA.

OECD (Directorate for Science, Technology and Innovation). 2018. *Bridging the digital gender divide include, upskill, innovate.* Geneva, Switzerland. www.oecd.org/going-digital/bridging-the-digital-gender-divide-key-messages.pdf. Accessed December 11, 2018.

Office of the Federal Register. 2016, December 19. Apprenticeship programs; Equal employment opportunity. *Federal Register.* www.federalregister.gov/documents/2016/12/19/2016-29910/apprenticeship-programs-equal-employment-opportunity. Accessed April 1, 2017.

Oldenziel, Ruth. 1999. *Making technology masculine: Men, women and modern machines in America, 1870–1945.* Amsterdam: Amsterdam University Press.

OpenNet Initiative. 2009. *Internet filtering in Egypt,* http://opennet.net/research/profiles/egypt. Accessed October 31, 2012.

Organization of Black Aerospace Professionals. 2018. *Captain Theresa M. Claiborne.* www.obap.org/index.php?option=com_content&view=article&id=294:theresa-claiborne&catid=36:pioneers-in-aviation-bios. Accessed January 8, 2019.

Osman, Ahmed Zaki. 2012, August 3. Women's movement: A look back, and forward. *Egypt Independent.* www.egyptindependent.com/womens-movement-look-back-and-forward/ Accessed February 15, 2013.

Ozer, Elizabeth M. and Albert Bandura. 1990. Mechanisms governing empowerment effects: A self-efficacy analysis. *Journal of Personality and Social Psychology,* 58(3): 472–486.

Paasonen, Susanna. 2005, January/June. Surfing the waves of feminism: Cyberfeminism and its others. *Labrys, feminist studies/estudos feministas/etudes feministes,* 1–14. Accessed January 5, 2010.

Pacey, Arnold. 1983. *The culture of technology.* Cambridge: MIT Press.

Parastoomarzieh blogspot. 2012. Free Parastoo and Marzieh. http://parastoomarzieh.blogspot.ca/2012/02/marzieh-rasouli-released-on-bail.html. Accessed October 9, 2012.

Parmar, Aradhana. Fall/2004. Ocean in a drop of water: Empowerment, water and women. *Canadian Woman Studies,* 23(1): 124–130.

Parra, Marcela Ossa, Roberto Gutiérrez, and María Fernanda Aldana. 2015, January. Engaging in critically reflective teaching: from theory to practice in pursuit of transformative learning. *Reflective Practice,* 16(1): 16–30.

Patterson, Anne W. 2013, May 21. The economic and political empowerment of women in Egypt: The way forward. Remarks to the American Chamber of Commerce in Egypt by the U.S. Ambassador to Egypt. www.amcham.org.eg/events-activities/events/843/. Accessed August 24, 2016.

PEOPLink. 2012, 2016. *PEOPLink.org website.* www.peoplink.org; http://www.peoplink.org/page/history. Accessed October 7, 2012 and again on June 15, 2016.

Pepper, Alexis. 2008, July 9. Extending the Olivia branch. *GOMAG.COM.* www.gomag.com/article/extending_the_olivia_bran/. Accessed August 8, 2010.

Pettinato, Michelle Sabolchick. 2013–2017. Leslie Ann Jones – Having the courage to raise your hand. *Sound Girls.org.* www.soundgirls.org/leslie-ann-jones/. Accessed April 10, 2017.

Pinch, Trevor J. and Wiebe E. Bijker. 1987. The social construction of facts and arti-facts: Or how the sociology of science and the sociology of technology might benefit each other. In *The social construction of technological systems: New directions in the sociology and history of technology*, edited by Wiebe E. Bijker, Thomas P. Hughes, and Trevor J. Pinch, 17–50. Cambridge, MA: MIT Press.

Plant, Sadie. 1995. On the matrix: Cyberfeminist simulations. In *Cultures of the Internet: Virtual spaces, real histories, living bodies*, edited by Rob Shields. London: Sage.

PLEN. 2018. *Women in STEM policy*. Preparing Women to Lead, Washington, DC. https://plen.org/stem/ Accessed December 15, 2018.

Pool, Robert. 1997. *Beyond engineering: How society shapes technology*. New York, Oxford: Oxford University Press.

Radio Farda. 2018, December 13. Rouhani order for more women in top posts 'ignored'. https://en.radiofarda.com/a/iran-rouhani-order-give-top-posts-women-ignored/29654374.html. Accessed December 13, 2018.

Radsch, Courtney C. 2012, May 17. *Unveiling the revolutionaries: cyberactivism and the role of women in the Arab uprisings*. Houston: James A. Baker Institute for Public Policy, Rice University. http://bakerinstitute.org/publications/ITP-pub-CyberactivismAndWomen-051712.pdf. Accessed October 19, 2012.

Ragai, Aziza and Ragai, Jehane. 2014. *Doria Shafik: A life dedicated to Egyptian Women*. Official website of Doria Shafik, Chapters XV-XXI. http://doria-shafik.com/doria-shafiq-feminism-egypt-activist.html. Accessed September 21, 2018.

Raghall, Karin. 2013, October 11. Unity around the demand for a quota with the women's movement in Egypt. *Kvinna till Kvinna and Women Living Under Muslim Laws*. www.wluml.org/ar/node/8748. Accessed August 24, 2016.

Ray, Nupur. 2014. Exploring 'empowerment' and 'agency' in Ronald Dworkin's theory of rights: A study of women's abortion rights in India. *Indian Journal of Gender Studies*, 21(2): 277–311. Accessed July 6, 2017.

Reagon, Bernice Johnson. 1976. Joanne Little. On 1976 vinyl record album Sweet Honey in the Rock, produced by Flying Fish Records, Chicago.

Rechtin, Mark. 2004. All-female Volvo gets mixed reviews. *Automotive News*, 78(6083), 8. Accessed September 6, 2010.

Reilly, Erin D., Kadie R. Rackley, and Germine H. Awad. 2017, April. Perceptions of male and female STEM aptitude: The moderating effect of benevolent and hostile sexism. *Journal of Career Development*, 44(2): 159–173.

Reporters Without Borders. 2012, March 12. Egypt chapter. *2011 Enemies of the Internet Report*, http://en.rsf.org/egypt-egypt-12-03-2012,42049.html. Accessed October 31, 2012.

Reporters Without Borders. 2018. *Violations of press freedom barometer: The figures in 2018*. https://rsf.org/en/barometer. Accessed December 13, 2018.

Rethink Breast Cancer. 2018, May 2. *Breast cancer statistics in Canada: The latest in breast health and breast cancer statistics in Canada*. https://rethinkbreastcancer.com/breast-cancer-statistics-canada/. Accessed August 29, 2018.

Richards, Pat. 2000. The spirit soars as the body moves and the mind discovers: Training for dragon boat racing for survivors of breast cancer. *Wave Link*, 10(4): 10. Publication of The Canadian Aquafitness Leaders Alliance Inc. www.calainc.org/downloads/wavelinks/Wavelink25.pdf. Accessed October 25, 2008.

Riley, Donna, Alice L. Pawley, Jessica Tucker, and George D. Catalano. 2009, Summer. Feminisms in engineering education: Transformative possibilities. *NWSA Journal*, 21(2): 21–40.

Roadwork Center. 2017. *Roadwork oral history and documentary project brochure. Building multi-racial coalitions through women's culture.* www.roadworkcenter.org/roadwork-oral-history-project/. Accessed May 14, 2017.

Rosenbrock, Howard H. ed. 1989. *Designing human-centred technology.* Berlin: Springer-Verlag.

Rosenbrock, Howard H. 1990. *Machines with a purpose.* Oxford & NY: Oxford University Press.

Rosendahl, Todd. 2013. Boden Sandstrom retirement. *Gender and Sexualities Task Force.* http://gstsem.pbworks.com/w/page/68357716/Boden%20Sandstrom%20Retirement

Rosser, Sue V. 2005. Women and ICT: Global issues and actions. In *Proceedings of the international symposium on Women and ICT: creating global transformation*, edited by Claudia Morrell and Jo Sanders. New York: Association for Computing Machinery.

Rotich, Juliana. 2012. *Afromusing blog: about Juliana Rotich.* http://afromusing.com/about/ and https://fr.twitter.com/afromusing. Accessed November 15, 2013.

Sandstrom, Boden C. 2002. *Performance, ritual and negotiation of identity in the Michigan Womyn's Music Festival.* (Unpublished Ph.D. Dissertation) University of Maryland.

Sarder, Russell, ed. 2016. *Building an innovative learning organization: A framework to build a smarter workforce, adapt to change, and drive growth.* Hoboken, New Jersey: John Wiley & Sons, Inc.

Schemm, Paul. 2010, June 13. Egypt café owner describes police beating death. *The San Diego Union Tribune.* www.sandiegouniontribune.com/sdut-egypt-cafe-owner-describes-police-beating-death-2010jun13-story.html. Accessed December 20, 2012.

Schwartzman, Roy and Merci Decker. 2008, Spring. A car of her own: Volvo's "your concept car" as a vehicle for feminism? *Studies in Popular Culture*, 30(2): 100–117. http://pcasacas.org/SiPC/30.2/Schwartzman_Decker.pdf

Scott, Ann, Lesley Semmens, and Lynette Willoughby. 2001. Women and the Internet: The natural history of a research project. In *Virtual gender: Technology, consumption and identity*, edited by E. Green and A. Adam, 3–27. London: Routledge.

Seeger, Peggy. 1971. *I'm gonna be an engineer.* www.peggyseeger.com/listen-buy/peggy-seeger-live/peggy-seeger-live-song-texts/im-gonna-be-an-engineer

Senge, Peter. 1990. *The fifth discipline: The art and practice of the learning organization.* New York: Doubleday.

Shah, Aditi. 2018, September 5. India soars above global average in hiring female airline pilots. *Reuters Business News.* www.reuters.com/article/us-india-pilots-women-analysis/india-soars-above-global-average-in-hiring-female-airline-pilots-idUSKCN1LL0OC. Accessed January 8, 2019.

Shalhoub-Kevorkian, Nadera. 2011, June. E-Resistance among Palestinian women: Coping in conflict-ridden areas. *Social Service Review*, 85(2): 179–204. Accessed October 20, 2012.

Sharafeldin, Marwa. 2012, March 15. The "hareem" of the new Egyptian Constitution. *Egypt Independent.* www.egyptindependent.com/hareem-new-egyptian-constitution/. Accessed August 14, 2012.

Shirazi, Farid. 2012. Information and communication technology and women empowerment in Iran. *Telematics and Informatics*, 29: 45–55. Accessed October 22, 2012.

Sinha, Saurabh. 2018, September 26. Number of women pilots in India doubles to 1,000 in 4 years. *The Times of India.* https://timesofindia.indiatimes.com/india/number-of-women-pilots-doubles-to-1000-in-4-yrs/articleshow/65957186.cms. Accessed January 8, 2019.

Skalli, Loubna H. 2006, Spring. Communicating gender in the public sphere: Women and information technologies in the MENA. *Journal of Middle East Women's Studies*, 2(2): 35–59.

Slack, Edward R. 2014, June 30. *Philippines under Spanish Rule, 1571–1898*. *Oxford Bibliographies.com*. www.oxfordbibliographies.com/view/document/obo-9780199766581/obo-9780199766581-0164.xml. Accessed March 22, 2017.

Smalley, Kristen. 2018, June 27. What's in store for Canadian STEM Professionals in 2018? *Ranstad*. www.randstad.ca/workforce360-trends/archives/whats-next-for-stem-jobs-in-canada_1751/. Accessed December 15, 2018.

Smith, Judy, ed. 1979. *Women and technology: Deciding what's appropriate*. Proceedings of a conference on women and technology, April 27–29, 1979. Missoula, MT: Women's Resource Center.

Smith, Judy, ed Judy. 1983. Women and appropriate technology: A feminist assessment. In *The technological woman: Interfacing with tomorrow*, edited by Jan Zimmerman, 65–70. New York: Praeger.

Smith, Judy and Ellen Balka. 1988. Chatting on a feminist computer network. In *Technology and women's voices: Keeping in touch*, edited by C. Kramarae. New York: Routledge & Kegan Paul.

Smithsonian National Air and Space Museum. 2015. *Women in aviation and space history*. https://airandspace.si.edu/explore-and-learn/topics/women-in-aviation/index.cfm. Accessed August 19, 2015.

Society for the History of Technology (SHOT). 2018. *Special interest groups*. www.historyoftechnology.org/special-interest-groups/. Accessed December 11, 2018.

Society of Naval Architects and Marine Engineers (Newfoundland Section). 2002, November 4. *SNAME news*. St. John's, Newfoundland, 1–7. www.engr.mun.ca/sname_rina/sname_news_2002.pdf

Songs of the Newfoundland Outports (Volume 1). 1965. Collected by Kenneth Peacock. National Museum of Canada Publication. www.wtv-zone.com/phyrst/audio/nfld/outports.htm. Accessed September 11, 2016.

Spender, Dale. 1995. *Nattering on the net: Women, power and cyberspace*. North Melbourne, Australia: Spinifex Press.

Stanley, Autumn. 1993. *Mothers and daughters of invention: Notes for a revised history of technology*. Metuchen, N.J: The Scarecrow Press.

Stash Records. 1978. *Women in jazz: All women groups*. Volume 1. Vinyl record album. Brooklyn, N.Y.

Staufenberg, Jess. 2016, April 28. Saudi Arabia is "not ready" for women drivers says deputy crown prince. *Independent*. www.independent.co.uk/news/world/middle-east/saudi-arabia-is-not-ready-for-women-drivers-says-deputy-crown-prince-mohammed-bin-salman-a7004611.html. Accessed July 10, 2016.

Stein, Jason. 2005. Designing women? They're still hard to find. *Automotive News*, 80 (6168), 2–30 H,30J. Retrieved from http://ezproxy.emich.edu/login?url=http://search.proquest.com/docview/219477727?accountid=10650

Stephan, Rita. 2013, August 28. Cyberfeminism and its political implications for women in the Arab World. *E-International Relations*. www.e-ir.info/2013/08/28/cyberfeminism-and-its-political-implications-for-women-in-the-arab-world/

Stepulevage, Linda. 2001. Becoming a technologist: Days in a girl's life. In *Virtual gender: Technology, consumption and identity*, edited by E. Green & A. Adam, 63–83. London: Routledge.

Step Up for Women. 1979. *Step Up for Women Program brochure*. Possessed by the author.

St. John's Women's Centre. 2018. *St. John's Status of Women Council*. https://sjwom enscentre.ca/. Accessed December 28, 2018.

Stoet, Gijsbert and David C. Geary. 2018, April. The gender-equality paradox in science, technology, engineering, and mathematics education. *Psychological Science*, 29(4): 581–593. Accessed December 15, 2018.

Styhre, Alexander, Maria Backman, and Sofia Börjesson. 2005. YCC: a gendered carnival? Project work at Volvo Cars. *Women in Management Review*, 20(2): 96–106.

Suchman, Lucy. 2009, March 5. *Agencies in technology design: Feminist reconfigurations*. Paper delivered at the 5th European Symposium on Gender & ICT - Digital Cultures: Participation - Empowerment – Diversity. University of Bremen, 1–15. www.informatik.uni-bremen.de/soteg/gict2009/proceedings/GICT2009_Suchman.pdf

Sullivan, Deana Stokes and Gillian Woodford. 2007, June 15. Cancer scandal puts path standards under the scope. *National Review of Medicine*, 4(11). www.nationalrevie wofmedicine.com/issue/2007/06_15/4_patients_practice03_11.html. Accessed November 19, 2008.

Sully. 2015, July 11. 'Sully' Sullenberg remembers the miracle on the Hudson. *Newsweek Special Edition*, Media Lab Publishing. www.newsweek.com/miracle-hudson-343489. Accessed September 17, 2015.

Talhami, Ghada Hashem. 1996. *The mobilization of Muslim women in Egypt*. Gainsville, FL: University Press of Florida.

Tewari, Ruhi. 2018, May 16. 'Moral cop' Muthalik & Sri Ram Sene make political debut but Karnataka says no thanks. *The Print*. https://theprint.in/politics/moral-cop -muthalik-sri-ram-sene-make-political-debut-but-karnataka-says-no-thanks/59333/. Accessed December 16, 2018.

The Telegram. 2008, September 15. Awakening the dragon at Octagon Pond. St. John's, NL. Print edition.

The Telegram. 2014, October 12. Avalon Dragons team finishes in top third at International festival. www.thetelegram.com/news/local/2014/10/27/avalon-dragons-team -finishes-in-top-thir-3917749.html. Accessed November 20, 2014.

Thomas, Robert J. 1994. *What machines can't do: politics and technology in the industrial enterprise*. Berkeley: University of California Press.

Tiburzi, Bonnie. 1984. *Takeoff! The story of America's first woman pilot*. New York: Crown Publishers.

Tilchen, Maida. 1984, June 23. A new wave in women's music . *Gay Community News*, 11(48): 7.

Torre, Lindsey A., Farhad Islami, Rebecca L. Siegel, Elizabeth M. Ward, and Ahmedin Jemal. 2017, April. Global cancer in women: Burden and trends. *Cancer Epidemiology, Biomarkers, and Prevention*. http://cebp.aacrjournals.org/content/26/4/444.full-text.pdf. Accessed August 29, 2018.

Tradeswomen, Inc. 2017. *Tradeswomen, Inc. website*. http://tradeswomen.org/. Accessed March 22, 2017.

Trescott, Martha Moore, ed. 1979. *Dynamos and virgins revisited: Women and technological change in history: An anthology*. Metuchen, NJ: The Scarecrow Press.

Trull, Teresa. 1977a. Woman-loving women. On 1977 vinyl record album The Ways a Woman Can Be produced by Olivia Records, Los Angeles.

Trull, Teresa. 1977b. Prove it on me blues. On vinyl record album Lesbian Concentrate produced by Olivia Records, Los Angeles.

Tucker, Sherrie. 1998, Fall. Nobody's sweethearts: Gender, race, jazz, and the Darlings of Rhythm. *American Music*, 16(3): 255–288.

Tufekci, Zeynep. 2017. *Twitter and tear gas: The power and fragility of networked protests*. New Haven, CT: Yale University Press.

Ulrich, Sheri. 2010. *Rolling River*. www.shariulrich.com/DSC10.php?offset=0&entry_id=2; www.shariulrich.com/discography/lyrics/Rolling%20River%20FINAL.doc. Accessed August 12, 2015.

UNESCO. 2011. *UIS statistics in brief. General profile – Egypt*. UNESCO Institute for Statistics. http://stats.uis.unesco.org/unesco/TableViewer/document.aspx?ReportId=124&IF_Language=en&BR_Country=2200. Accessed October 31, 2012.

UNESCO. 2012, May 13. *Campaign for Literacy and the Renaissance of Egypt: 2012–2020, press release*. UNESCO Office in Cairo. www.unesco.org/new/en/cairo/about-this-office/single-view/news/campaign_for_literacy_and_the_renaissance_of_egypt_2012_2020/. Accessed October 31, 2012.

Ushahidi. 2008–2016. *About Ushahidi*. www.ushahidi.com/about. Accessed July 10, 2016.

Wahba, Dina. 2016. Gendering the Egyptian Revolution. In *Women's movements in post-"Arab Spring" North Africa*, edited by Fatima Sadiqi. New York: Palgrave Macmillan.

Wajcman, Judy. 1991. *Feminism confronts technology*. University Park, PA: The Pennsylvania University Press.

Wajcman, Judy. 2004. *TechnoFeminism*. Cambridge, UK: Polity Press.

Wajcman, Judy. 2007. From women and technology to gendered technoscience. *Information, Communication & Society*, 10(3): 287–298.

Wakeford, Nina. 1997. Networking women and grrrls with information/communication technology: Surfing tales of the world wide web. In *Processed lives: Gender and technology in everyday life*, edited by J. Terry and M. Calvert, 52–66. London: New York.

Watkins, Karen E. and Virginia J. Marsick. 1992. Toward a theory of informal and incidental learning in organizations. *International Journal of Lifelong Education*, 24(4): 287–300.

Watkins, Mary. 1977. Don't pray for me. On 1977 vinyl record album Lesbian Concentrate produced by Olivia Records, Los Angeles.

Waves of Hope. 2005. Breast cancer survivor dragon boat teams. *Waves of Hope website*. www.wavesofhope.ca/content/Teamlinks. Accessed October 26, 2008.

Wayne State University. 1969. *WSU Student Organization Rosters. 1960–69*. Walter P. Reuther Library, Wayne State University Archives. https://reuther.wayne.edu/files/Student_Organization_Rosters_1960-1969.pdf

Wayne State University. 1970. *WSU Student Organization Rosters. 1970–79*. Walter P. Reuther Archives, Wayne State University Archives. https://reuther.wayne.edu/files/Student_Organization_Rosters_1970-1979.pdf

Webster, Andrew. 1991. *Science, technology and society*. New Brunswick: Rutgers University Press.

Wertheimer, Barbara Mayer. 1977. *We were there: The story of working women in America*. New York: Pantheon Books.

Whitelaw, John P. 2008, July/August. Currents: Offcuts. *Wooden Boat*, (203): 25–27.

Wider Opportunities for Women. 2017. *Wider Opportunities for Women website: Our story*. www.wowonline.org/our-story/. Accessed March 22, 2017.

Williams, Christopher. 2011, January 28. How Egypt shut down the Internet. *The Telegraph*. www.telegraph.co.uk/news/worldnews/africaandindianocean/egypt/8288163/How-Egypt-shut-down-the-internet.html. Accessed October 31, 2012.

Williams, Zoe. 2017, October 16. Sexual harassment 101: What everyone needs to know. *The Guardian*. www.theguardian.com/world/2017/oct/16/facts-sexual-harassment-workpla ce-harvey-weinstein. Accessed December 13, 2018.

Williamson, Chris. 1975. Song of the soul. On 1975 vinyl record album The Changer and the Changed produced by Olivia Records, Los Angeles.

Wise, Jeff. 2011, December. 6. What really happened aboard Air France Flight 447? *The Best of Popular Mechanics*. www.popularmechanics.com/flight/a3115/what-really-happened-aboard-air-france-447-6611877/. Accessed September 17, 2015.

Women in Aviation International (WAI). 2015. *Some notable women in aviation history*. www.wai.org/resources/history.cfm. Accessed August 23, 2015.

Women in Aviation International (WAI). 2016. *Lt. Cmdr. Brenda E. Robinson*. www.wa i.org/pioneers/2016/lt-cmdr-brenda-e-robinson. Accessed May 20, 2018.

Wright, Tessa. 2016, May. Women's experience of workplace interactions in male-dominated work: The intersections of gender, sexuality and occupational group. *Gender, Work & Organization*, 23(3): 348–362.

Yousef, Hoda. 2011, Winter. Malak Hifni Nasif: Negotiations of a feminist agenda between the European and the colonial. *Journal of Middle East Women's Studies*, 7(1): 70–89, 131. Accessed November 1, 2012.

Zakhary, Dalia. 2016, March 8. Women's rights in Egypt. *International Idea News*. International Institute for Democracy and Electoral Assistance (IDEA) News. www.idea.int/ news-media/news/women%E2%80%99s-rights-%C2%A0egypt. Accessed August 24, 2016.

Zeid, Amir and Fatima Al-Khalaf. 2012, September. The role of social media in the Egyptian Revolution: The initiation phase (2010–2011). *American Academic & Scholarly Research Journal, Special Issue*, 4(5): 1–10. www.naturalspublishing.com/files/p ublished/g8442e995whq19.pdf

Zeidan, Sami. 2010. *Navigating international rights and local politics: sexuality governance in a post-colonial setting* (Ph.D. dissertation). Political Science, The City University of New York.

Zuberi, Hena. 2011, May 22. The Rosa Parks of Saudi Arabia: Women challenging the ban by driving. *MuslimMatters.org*. http://muslimmatters.org/2011/05/22/the-rosa-parks-of-sa udi-arabia-women-challenging-the-ban-by-driving/. Accessed October 25, 2012.

Index

ABC Records 83, 95
Abdullah, King 11
abortion 56, 101
Actor Network Theory 4
Adam, M. 81
adult education 13–14, 159
adult learning theory 2–4
Africa 46, 51
African-Americans 2, 80, 110, 117–19
Afromusing 10
age gaps 61
agency vi, 4–5, 10
Ahmed, L. 48
Air Florida 131
Air France 136, 139
Air Race Classic 147
Air Traffic Controllers 131, 135–6, 152
Airline Pilots Association 119
Alaska 127
Alive 81
Alliance of Arab Women 73
American Airlines 119, 136
American Federation of State, County and
    Municipal Employees (AFSCME) 110
American University 98
American University in Cairo 61
analog audio systems 84–5
Andres, N.J.L. 108–15, 152–3, 156–7
Andres, P. 108
Andres, R. 108
Angel Flight 133
Anonsen, A.M. 18–20, 25–34, 38, 41,
    153–4, 159
anthropology 10, 141
apprenticeships 93–8, 107, 114, 153
Arab Spring 45, 57
Arab states 45–6, 50–1
Arab Women Organization (AWO) 51
Arabic language 12, 46, 51

archive systems 46
Arizpe, L. 10
Armstrong, F. 80
Asia 16, 122
Asian-Americans 2, 80
Asian-Pacific Islanders 108, 115
Asset Pro 85
Associated Press (AP) 57
Atlantic 118, 147
audio engineers 2, 5, 14, 82–8, 90–9,
    101–6, 150–4, 156–9
Australia 15, 17, 44, 118
authoritarianism 49–50, 57
automation dependency 136–7
automotive industry 8–9, 79–80
Avalon Dragon Boating Association
    15–44, 150–1
aviation 2, 14, 117–49, 152, 156
Aviation Flight Technology 121
Aviation Management 121
Awad, G.H. 158

Backman, M. 8
Badcock, C. 34
Badcock, H. 34
Badran, H. 72
Badran, M. 48–9
Bahamas 119
Bahrain 45
Balka, E. 7
Baltimore Community Downtown Sailing
    Center 157
Bandura, A. 4

ElBaradei, M. 68

Bare Naked Ladies 83
Barshi, I. 138
Beese, M. 117

benefits of technology mastery 2, 9, 13–14, 18–19, 110; and audio engineers 80–1, 101–6; and boatbuilders 34, 38–43; and cyberactivists 48, 51–4, 69–71, 78; and empowerment 152, 156–7, 159; and pilots 122, 143, 145–9, *see also* technology mastery
Berklee College of Music 92, 94
Berlin Wall 63
Bettney, J. 16–19, 21, 25, 35, 37, 159
Bhutto, B. 63
Birchler, J. 145
Bishop Spencer College 24
Blackberry 71
blogs 10–11, 47, 53–4, 60, 145
blues music 80–1
boatbuilding 15–44, 152–4, 156–9
Börjesson, S. 8
Boserup, E. 6
Brandy, C. 81
Brazil, A. 18, 20–1, 24–5, 28, 40–1
breast cancer 2, 15–44, 150–1, 154
Britain 7, 17, 49
British Columbia Cancer Institute 17
Brown, Jane 18–21, 23–8, 30, 32, 34, 36, 38, 41, 150–2
Brown, Jill 119
Bush, C.G. 7
Bussey, D. 81

Cadillac Motor Cars 114
Cairo University 50–1, 63
Campaign to Change the Male Face of Parliament 11
Canada 1–2, 15, 17, 20, 44, 83, 126
Canadian Breast Cancer Foundation 16–17
Canadian Cancer Society 19
Caradonna, E. 114
career paths 2, 24–5, 61–4, 109–10, 115–16; and audio engineers 82–3, 86–98, 101, 103–4; and cyberactivists 71, 78; and empowerment 151, 156–7; and pilots 120–32, 141–2, 144, 146, 148–9
Caribbean 126, 130
Carter, J. 107
Carter, M. 80
case studies 2, 4, 13–14, 45, 50, 107, 120, 127, 150–9
Casner, S.M. 138
Catholics 24, 108
cell phones 46–7, 97
censorship 11, 46–7, 155

Central Airlines 119
Central America 130, 142
certification 2, 115, 120, 126, 128, 138, 142
Challenger Space Shuttle 129
Chaney, E.M. 6
Chapman, T. 83
chemotherapy 27, 34, 43
Chesapeake Bay 128
Chicago Women in Trades 107
childhood/children 1, 108–9, 114, 119–20, 123–4; and audio engineers 86, 88, 91–2; and boatbuilders 21–3, 25, 34, 42; and cyberactivists 48, 63, 66, 74–6; and empowerment 150–1, 156–7; and pilots 127, 130, 133, 136, 140–2, 146–7, 149
China 16, 77, 115
Christian, M. 81, 97
Church of England 24
citizenship 73–4
Civil Air Patrol 120
civil service 110
civil society 64, 69
Claiborne, T.M. 119
Clarke, P. 119
class structure 2, 10, 12, 47, 59, 65, 82, 155
Clemmens, G. 81
CNN 57
co-construction thesis 6
Coalition of Labor Union Women (CLUW) 107
Cochran, J. 118
Coleman, B. 117–18
colonialism 6, 17, 47, 49, 79
commercial airline pilots 2, 14, 117–54, 156–8
Commission on the Advancement of Women 8
Committee to Protect Journalists (CPJ) 47, 77
communication 2, 9, 12, 46, 51–2, 58, 60, 65, 135–6, 139–40, 149, 154
Comprehensive Employment and Training Act (CETA) 110
computers 1, 8–9, 33, 54, 57–63, 85, 88, 133–4, 136–8, 151
Computerus 9
concerts 82–3, 86, 90, 101, 106
confidence 18, 141, 145–6, 151, 157; and audio engineers 98–101, 106; and boatbuilders 20, 24–6, 29, 33, 38–40, 43; and cyberactivists 58–62, 66, 68–9; and pilots 118, 132, 137; and sewer worker 109, 114
connectivity 54

conservatives 66–7, 74
Consortium of Pub-going, Loose and Forward Women 10
Constituent Assembly 73–4
contextual factors 2, 5–7, 9, 12–13, 151–2, 158–9
Convention on the Elimination of All Forms of Discrimination against Women (CEDAW) 76
Cook, B.J. 159
cooperatives 79
corporations 55, 79–80, 87, 126, 143
corruption 45
Cotten, E. 80
Cox, I. 81
craftswomanship 35
Crew Resource Management (CRM) 139–40, 149
critical reflection 3–4
Culver, C. 81, 90–1, 97
curfews 66, 89
Curran, H.P. 107
cyberactivists 9–10, 13, 45–78, 150, 154–6
Cyrillic 115

Dabinett, D. 36
Daoud, L. 77
Darlings of Rhythm 80
databases 9
Davis, C. 98
Davis, D. 98
DC United Soccer 104
dead reckoning 138
Decker, M. 8
Dell Purse Book 79
Delta Zeta sorority 109
democracy 11, 45, 47, 50, 63, 69, 73–8, 140, 150, 159
Department of the Army 127
Desert Storm 129–30
determinism 156
Detroit General Hospital (DGH) 110–11, 116, 152
Detroit Medical Center 110
Detroit Water and Sewerage Department 2, 14, 107–16, 150–3, 156–8
developed countries 10
developing countries 10
development agencies 6, 50
dial-up modems 46–7, 59
Dickens, H. 81
digital audio systems 84, 85
digital cockpit systems 123, 148
digital divides 46

Digital Performer 85
Disability Determination Examiner 115
disc jockeys 97
discrimination 12, 65–6, 73, 75–6, 120, 144–5
divisions of labor 6, 18, 25, 83, 152, 156
Dlugacz, J. 90
Dobkin, A. 81
Dodge Main 109–10
dragon boat racing 2, 5, 14–44, 153–4, 156–9
Drydyk, J. 5
dualism 8, 75
Dubai 72
Duley, Margaret 17
Duley, Margot I. 17–18
Dulles Airport 128

e-mail 45, 51, 56, 58–9, 61, 153
e-resistance 12
Eagles Mere Air Museum 146
Earhart, A. 118
East Germany 63
Eastern Michigan University 88
Ebo, B. 9
Eccles, J. 1
education 1–4, 10, 13–14, 24, 46–7; and audio engineers 88, 95, 106; and cyberactivists 49–51, 58–60, 64, 66–7, 69–70, 78; and empowerment 150–1, 153–4, 156, 158–9; and pilots 120, 124, 126–7, 143, 145; and sewer workers 107–10, 115–16
Edwards, H. 110
Egypt 2, 5, 10, 13–14, 45–78, 150–1, 153–4, 156–9
Egyptian Center for Women's Rights 76
Egyptian Constitutions 45, 67–8, 71–6
Egyptian Feminist Union 47, 68, 72–3
Egyptian Ministry of Communications and Information Technology 46
Egyptian Revolution 2, 10, 14, 45–78, 151, 158
Egyptian Universities Network 46
elections 10–11, 50, 59, 71–3, 76–7
Elliott, M. 18, 20, 23, 28, 31, 33–4, 38–9
Elliott, N. 18, 27, 29, 33–4, 36–8
Eltahawy, M. 47
Embry Riddle Aeronautical University 119, 139
Emergency Medical Technicians 109–10
emigration 6
empowerment/empowerment theory 2, 4–5, 7, 9, 12–14, 66, 70, 76, 102, 150–9

engineering 1, 8, 63–4, 80, 82–93, 119, 129
England 16, 44, 62, 84, 117–18
English Channel 118
English language 50–1, 115
English, L.M. 158
equality 2, 7–8, 12, 47, 49–51, 57, 69,
    71–3, 75–8, 107, 150, 158–9
equalization (EQ) 84
ergonomics 7
Esser, D. 139
ethnomusicology 83, 104
Europe 6–7, 16, 130
European Parliament 76
Everett, A. 9, 155
Executive Orders 107
Expectancy-Value Model of
    Achievement-Related Choices 1
Exploring Diversity in Technology's
    History (EDITH) 6

face-to-face organizing 57–8, 78, 154–5
Facebook 10–11, 47, 51–61, 71–2, 154–5
family 1–2, 12–13, 17–18, 20–1, 23–5; and
    audio engineers 86, 91–2; and boatbuilders
    33–4, 40–2, 44; and cyberactivists 50, 52,
    56, 58–60, 64–5, 67, 74, 76–7; and
    empowerment 150–2, 156–7; and pilots
    120, 122–3, 125, 127, 130–1, 135, 146, 149;
    and sewer workers 108, 115–16
Faulkner, W. 8, 159
fax machines 47
Federal Aviation Administration (FAA)
    120–1, 130, 138, 143, 152
Feldman, M. 81
feminism/feminists 1–2, 4–9, 11, 16–17,
    80–2; and audio engineers 88–9, 93, 97,
    100–1, 104, 106, 152, 157; and Egypt
    47–50, 59, 64, 67–69, 74, 151, 154; and
    Islam 47–8; cyberfeminists 9–10, 45–78;
    and empowerment 150–9; feminist
    technoscience 6; and Newfoundland
    159; and pilots 119; role of 158–9; and
    skilled trades 25, 107
feminist technology assessment 7–8
feminist technology values 7
festivals 80, 82–3, 86, 88–9, 91, 101–4, 156
First Ladies 51
First World War 36, 118
fishing industry 17–18, 22–3, 33
flight instructors 123–4, 126, 129, 131, 139,
    142–3
Flood, R. 33, 39
Flood, S. 18–20, 24–6, 29, 31, 33, 39, 41
Flower Ceremony 43–4

Flower, R. 81
folk music 80–1, 83, 88, 91, 102, 105–6
Fox Hollow Folk Festival 80
France 46, 117–18
Free Internet initiative 46
free speech 55, 76
freelance work 88–9, 95, 97, 104
Freire, P. 4
frequency spectrum 84
Frontier Airlines 119
Fuchs, L. 82
Furies Collective 100

Gardner, K. 83
gender audit assessment methodology 8
gender bias 6, 8, 10, 142–4
gender role stereotypes 2, 6, 8–9, 64–5, 67,
    80, 125, 130, 141
gender studies, 51, 64
George Peabody College 87
George Washington University 90
Germany 51, 63, 117, 147
Gezi Park 78, 155

El-Ghobashy, M. 75

Ghonim, W. 72
Girl Guides 42
glass ceilings 101, 119
Global Positioning System (GPS)
    133–4, 138
Global System for Mobile Communication
    (GSM) 46
Goldenrod Music 82
Google 72, 126
Gordon, L. 88
governance 45, 78
gradualism 49
Grammy Award 83
Grand Ole Opry 87
Grant, T. 82
Great White North 35
Green, E. 7
Greene, P. 18, 23–6, 32, 36–7, 40

Habitat for Humanity 38
Hackler, A. 102
hactivism 9, 55
Hafter, D. 6
ham radios 47
hand tools 20–1, 31–2, 151, 153
handicrafts 61–2
Haneda Airport 122
Hanley Audio Systems 89

harassment 12, 47, 50, 55–6, 67, 70–1, 78, 154, 156
Haraway, D. 9
hardware 54, 136
Harris Institute 95
Harris, S. 15, 19
Hassan, M. 76–7
hate speech 52
Hatem, M.F. 49
Hawaii 123
health care 7, 12, 27–8, 39–40, 44, 49, 79, 159
hegemony 155
Helem 12
Hewlett, H. 117
Higbie, B. 81
Hinduism 10
Hispanic Americans 80
HIV/AIDS 12, 115–16
Hong Kong 16
Horowitz, A. 101
house mixers 83
housework 25, 31, 33, 36, 66, 74, 141
Howell, D. 18, 20–3, 26, 28–30, 32, 39, 42–3, 151
Hudson, J.L. Jr. 109
Hudson River 137
human rights 11–12, 47, 56, 76–7, 158
Human Rights Watch 76
human-centered systems design 7
Hunt, D. 88

ID cards 73
identity theft 55
India 10, 119
individual candidacy (IC) 73
industrial arts 24, 42
industrialization 6, 13
informal learning 2–4, 13–14, 25–30, 77, 111–13, 116, 138, 153, 158
information and communications technologies (ICT) 2, 9–11, 13, 45–78, 155, 157
Institute for the Musical Arts 102
Intermedia Sound 94
International Alliance of Women 75
International Breast Cancer Paddlers' Commission Participatory Dragon Boat Festival 44
International Dragon Boat Festival 16
international law 50–1
International Sweethearts of Rhyme 80
International Women's Day 71
Internet 9–13, 45–7, 50–1, 53–4, 56, 58–9, 61, 71–2, 155–6

Internet Service Providers (ISPs) 46–7
internships 64, 67, 110, 122–3
intersectionality 6
Intifada 63
inventors 8
iPad 134, 138
Iran 11, 51, 59, 67
Iraq 51
Ireland 44
Islam 47–8, 56, 68, 71, 73–5
Israel 12, 63, 96
Italy 16, 44

Jackson, G. 112–13, 116
Japan 122, 130–2, 146
Japanese language 122
jazz music 80, 83, 89
Johnson, A. 118
Johnson, D.G. 6
Johnson Geo Centre 16–17, 19
Johnson, L.B. 107
Jones, L.A. 82–3, 86, 92–3, 95–7, 99, 102–4
journalists 11, 50–1, 53, 58, 62–4, 67, 77
journals 11, 155
Juno Award 83

K-12 1
Kamel, B. 47
Kane, K. 82–3, 85, 91–2, 95–100, 102–3, 150
Karimi, S. 11
Karnataka Legislative Assembly 11
Kazakhstan 114–15, 157
Kefaya 58
Kennedy Center 82, 104
Kenya 10
Kidney Foundation 116
Koth, K. 81
Kurds 78, 155

labor movement 109
labor studies 5–6, 13–14
Ladyslipper Music 82
Lafayette Clinic 110
Laing, K. 19
Lammas Bookstore 101
Lansing Community College 107
Laroche, R. de 117
Latin America 6
League of Arab States 51
learning organizations 3
Lebanese Penal Code 12
Lebanon 12–13, 63, 77

Lebow, E.F. 117–18
Lems, K. 81
Leslie University 122
LGBTQ issues 12–13, 81–3, 90, 100–1, 103, 110, 125, 144, 155
liberals 47, 66, 74–5
Libya 45
licenses 79, 117–18, 124, 128–9, 131, 138
Lifeline 81
Lilith 81
Lim, P. 108
Lindroth, J. 120, 123–4, 132–4, 136–8, 141–8
line-oriented flight training (LOFT) 138
literacy 46, 57
Lithicom, T. 97
Lithuania 114–15
Logic 85
Lohan, M. 6
Lowell Institute 98
Lukenson, G. 97
lymphedema 15, 19, 26–8
lyrics 81

Macdonald Drive Junior High School 18, 27, 34, 36, 42
McGuire Air Force Base 129
Macintosh 85
McKenzie, D. 15
Mackey, J. 81
McLarney, E. 75
Madsen, S.R. 159
Mahfouz, A. 47
Malaysia 16
male domination 8–9, 80, 93–101, 102, 105, 107, 116, 120, 156
Manx, H. 83
Mårtensson, L. 138
Martin Luther King Jr. Memorial Library 90
mathematics 1, 21, 24, 63, 87–9, 126–7, 132
Mauritania 51
media 11, 16, 20, 37, 50–1, 53–4, 57, 62, 64, 77, 156
Meem 13
Memorial University Marine Institute 17
mentoring 1, 3, 13, 61, 63; and audio engineers 86, 94–8, 102, 106; and boatbuilders 24, 26, 42, 44; and empowerment 151–2, 154, 157; and pilots 126, 130–2, 149
Meyer, A. 120
Mezirow, J. 3–4, 153–4
Michigan Department of Education 115
Michigan Department of Labor 107, 115

Michigan Rehabilitation Association 116
Michigan Rehabilitation Services 115
Michigan Women's Commission 107
Michigan Womyn's Music Festival 82
Middle East and North Africa (MENA) 11, 46, 51
military 55, 72, 74, 76–7, 118, 120, 124–5, 129–30, 140–1, 143
Millington, J. 81, 102–3
Million Man March (MMM) 82
Million Women March (MWM) 155
minorities 1–2, 12, 74, 156–7
misogyny 9
mobile phones 46, 52–3, 71
Moisant, M. 117–18
Mojab, S. 78, 155
Mongolia 115, 157
monitor engineers 83, 85
Moral Majority 9
Morgall, J. 7
Morocco 46, 51
Morse Code 132
Morsi, M. 55, 71, 75–6
mosques 48
Mostafa, D. 77
motivation 1–2, 4, 13–14, 111, 116; and audio engineers 86–93, 99, 101–3; and boatbuilders 18–20, 40, 44; and cyberactivists 62–4, 77; and empowerment 150–2, 157, 159; and pilots 117, 120–32, 147, 149
Mount St Vincent School 24
Mubarak, H. 14, 45, 47, 55, 68, 72, 74, 76
Mudge, G.D. 79
Muhammad, Prophet 11
multitasking 135, 141
Musa, N. 47, 50
music/music industry 79–106, 150–2, 154, 157–9
musical instruments 81, 86, 91
musicians 80–1, 87, 91, 100, 103–4, 152, 154
Muslim Brotherhood 74
Muslims 11, 48, 74
Muthalik, P. 10–11

Nabrawy, S. 50
Nasawiya 12
Nasif, M.H. 47, 49
Nasser, G. 49, 72
Nation Airlines 122
National Apprenticeship Act 107
National Campaign for Literacy 46
National Organization for Women 82

National Sound 97–8
National Women's Mailing List (NWML) 9
National Women's Music Festival
  (NWMF) 80, 82, 103
nationalism 47–9, 155
Native Americans 80
Naval Architecture Diploma Program 17
navigation 132–6, 138, 149–51, 153–4
Nazra for Feminist Studies 76–7
Near, H. 81, 106
neo-liberalism 74
Network for Women's Rights
  Organizations 73
networking 4, 40, 70, 73, 82, 93, 109, 149,
  152–3, 155–8
New Orleans Jazz and Heritage Festival 89
New World Fitness 18–20, 34
New Zealand 15, 44
Newfoundland 14, 16–20, 29, 36, 42, 150,
  152, 159
Newfoundland and Labrador House of
  Assembly 16
Newfoundland and Labrador Road Build-
  ers/Civil Association 37
Newfoundland Regiment 36
Nile, River 45
Ninety-Nines 118, 120–1, 126, 130, 145,
  147, 149, 152
non-governmental organizations (NGOs)
  8, 50, 58, 64, 77
Nontraditional Employment for Women 107
North America 34
North American-Soviet Tradeswomen
  Exchange for Peace 114–15, 152
Northeast Detroit Water Treatment
  Plant 110
Northeast Institute 98
Nutter, C. 139

obstacles 24, 54–5, 64–7, 78, 120; and
  audio engineers 87, 89, 92, 96, 98; and
  empowerment 150, 156, 158; and pilots
  141–5, 149
Octagon Pond 37, 39, 43
Odyssey Yachts 35–6
Office of Women and Work 107
Okolloh, O. 10
Olivia Records 81
Olivia Music Collective 90
Olympic Games 15
on-the job training (OJT) 93–4, 97, 106,
  139, 149, 153
Open University 7
open-source software 10

Operation Desert Shield 129
oppression 4, 64
Orange County Airport 143
Organization of Black Aerospace
  Professionals 119
Orientation to Trades and Technology 38
Ouellette, D. 18, 20, 34, 40
Ouellette, J. 18, 20, 23, 30–1, 33–4, 38, 40
Our Bodies Ourselves 79
outport 22
Owen, J. 7

Paddle in Paradise Festival 43
Pain, D. 7
Pakistan 51
Palestinians 12, 51
paradigm of empowerment 150–9
parents 24, 41, 51, 61–2, 108; and audio
  engineers 86, 88–9, 92, 106; and
  cyberactivists 64–5, 67; and
  empowerment 151; and pilots 121,
  123–4, 127–8, 130–1, 142, 146
Parlio 72
Parmar, A. 5
paternalism 9
patriarchy 2, 4, 47–9, 71, 155, 159
patriotism 74
Patton, J. 120–3, 130–2, 134–5, 138, 140,
  144–5, 147–52
peer learning 29, 44, 153
People's Assembly (PA) 72–3
People's Councils 49
petitions 52
Pharaohs 45
Philippines 8, 108
physiotherapy 15, 21, 27
pilot training 138–41, 149
pilotage 138
pilots 2, 14, 117–54, 156–8
Pink Chaddi Campaign 10
pioneers 2, 18, 44, 82, 103, 117–19, 143,
  152, 158
Plant, S. 9
Pleiades Records 81
plug-ins 85
pluralism 75
Poland 16
police 55, 68, 71–2, 77
political science 51, 62–4
pollution 79
post-colonialism 47
Potomac River 131
poverty 111
Powder Puff Derby 147

Powell, B. 119
power relations 6, 9, 13
power tools 16, 21–2, 27, 31, 39, 111, 153
PowerPoint 59
prejudice 141–5
Presley, E. 80
Prezi 59
prison 55
Pro Tools® 85, 95
professionalism 102–3
Project Hope 116
proportional representation (PR) 72
psychosocial factors 1–2, 13
public sphere 11, 159

quality 28, 34, 44
Quidi Vidi Lake 36–7
The Quilt 30
Quimby, H. 117–18
Quran 48, 73

racism 65, 79, 117–18, 120, 140, 155
Rackley, K.R. 158
Rainey, G. 81
Ray, N. 5
Reagon, B.J. 81, 101
recessions 109, 111
record companies 81, 92, 95
Recording Institute of America 83, 95
Recording Workshop 98
Redskins 104
Redwood Records 81
Reel World String Band 81
Rehabilitation Counselor 115–16
rehabilitation sciences 15, 21
Reilly, E.D. 158
Renaissance of Egypt 46
repair shops 79
repression 45–7, 55, 76
research questions 1–2, 13, 18, 50, 150–9
respect 112–14, 116
Reuters 57
Reynolds, M. 81
Rich Site Summary (RSS) 57
Richey, H. 119
Roadwork 101
Robert F. Kennedy (RFK) Stadium 82, 104
Robinson, B. 119
rock-and-roll music 80, 91
Rogers, G. 16, 19, 159
role models 1, 61, 63, 88, 92, 106, 125, 149, 157
Rose, B. 81
Rotich, J. 10

Rouhani, H. 11
Royal Institute of Technology 138
Runstein, R. 94
Rush 97
Russia 88, 117, 127

el Saadawi, N. 50, 57, 64

Sadat, A. 49
Saeed, K. 72
safety 20, 26–7, 39, 58, 64–5, 67, 113–14, 121, 127, 139–40, 152
Said, K. 53
St Lawrence University 89
Salafists 74
Sandler, R. 107
Sandstrom, B. 82–3, 86, 89–91, 97–8, 100–2, 104, 152
Saudi Arabia 11, 45
Schmink, E. 6
School of Oriental and African Studies (SOAS) 51
Schumacher, E.F. 7
Schwartzman, R. 8
science, technology and society (STS) studies 5–6, 13–14, 156, 159
Second World War 132, 144, 147
Seeger, P. 80
segregation 24–5, 48–9
Senge, P. 3
seniority 121, 142, 146
September 11 2001 121
sermons 48
sewer maintenance 2
Sex Role Impact Statement (SRIS) 7
sexism 3, 12, 67, 79, 89, 92, 105, 120, 140, 155
sexual harassment 12, 50, 67, 71, 78, 154, 156
sexually transmitted infections (STIs) 12
Shafik, D. 50, 72
Shalhoub-Kevorkian, N. 12
Sharafeldin, M. 73–4
Sha'rawi, H. 47–9, 68
Sharia Law 51, 73
al-Sharif, M. 11
Siciliano, T. 120, 127–30, 132–3, 135–6, 138–9, 142, 146, 151–3
simulators 138
Singapore 44
singers 81, 104, 106
al-Sisi, A.F. 76–7
Skype 13, 50, 61
smart phones 54

Smith, J. 7
Smithsonian Institution 82, 91
social constructivism 5–6
social justice 69, 78, 101, 116, 158
social media 11, 45–78, 151, 153–4
social sciences 51, 125, 132
socialism 101
Society for the History of Technology
    (SHOT) 6
socio-technical systems theory/socio-techni-
    cal skills 135, 149, 153, 156, 158
software 10, 85, 128, 136
solidarity 34, 68–9
Sonar 85
songwriters 81, 106
South Africa 46
South America 130, 142
spam 55
spirituality 148–9
sports medicine 15
Sri Ram Sene 10–11
STEM education 1, 158
STEM Initiative 157
STEM subjects 1, 8, 157–8
Step Up for Women 107
Stepulevage, L. 150
stereotypes 1–2, 6, 64–5, 67, 80, 110, 130,
    142, 145
Stinson, K. 118
Styhre, A. 8
Suchman, L. 4
suffrage 17
Sullenberg, C. 137
Supreme Council of the Armed Forces
    (SCAF) 45, 47, 55, 72
surveillance 46, 55
Swahili language 10
Sweden 138
Synergetic Audio Concepts 98
Syria 45

Tahrir Square 58, 63, 71
Taksim Square 78, 155
Talhami, G.H. 49–50
Teaching Artists to Reach Technological
    Savvy (TARTS) 9
technology mastery 2, 4, 13–14, 51–2, 54;
    and appropriate technology movement 7;
    and audio engineers 79–93, 102, 106; and
    boatbuilders 30–1, 38; and cyberactivists
    58–62, 77; and empowerment 150–9;
    implications for 157–9; and pilots 120–1,
    125, 132–8, 141, 149, *see also* benefits of
    technology mastery

technology limitations 13, 54, 57, 78, 149, 154
technomodernism 155
television 53, 63, 71, 83, 123, 131
Texas International Air 119
text messaging 46, 58
Thabet, M. 50
Third Technology for the People Fair 8
Thornton, W.M. 80
Tian Shan Mountains 115
Tiburzi, B. 119
Tillery, L. 81
torture 76
tourism 16, 147
Tradeswomen Inc 107
trailblazers 106, 117
transferable skills 29–30
transformative learning theory 3–4, 153–4
transnational companies 6
Trevor, M. 81
trial and error 77, 97, 106, 153
Trull, T. 81
Tufekci, Z. 78, 155
Tunisia 45, 51
Turkey 77
Twitter 10, 12–13, 47, 51–6, 59–61, 71,
    154–5
Tyson, W. 90

Uganda 6
Ukraine 114
Ulrich, S. 44
unemployment 109–10
UNESCO 46
UNIFEM 64
Union of Soviet Socialist Republics
    (USSR) 114–15, 152
unions 57, 95, 107, 109–10, 119
United Airlines 119
United Nations (UN) 46, 64, 75
United Parcel Service (UPS) 119
United States Air Force (USAF) 120, 123,
    125, 129, 150
United States (US) 2, 8–9, 14, 16, 23; Air
    National Guard 120; and audio engi-
    neers 79, 96; and boatbuilders 44; and
    cyberactivists 59, 67, 69, 72, 75–6; and
    empowerment 159; and pilots 117,
    119–20, 122, 126, 132, 138, 140, 146;
    and sewer workers 108, 110, 114
University of British Columbia 15
University of London 51
University of Maryland 83, 104
University of Massachusetts (U-Mass) 88
University of Michigan 90

University of North Carolina 83
Ushahidi 10

Vandenburgh, W. 136
Vater, B.A. 16
Very High Frequency Omni-Directional Radio (VORs) 132–3
Vietnam War 79, 89, 101, 109–10, 130
violence 9–10, 12, 50, 52, 56, 68, 72, 76
Vogl, N. 81
Volvo 8

Wahba, D. 158
Wajcman, J. 6–7
Warner, E.H. 119
Warner Theater 104
Washington, R.G. 119
Water Utility Workers in Maintenance and Repair 107, 110–14
Watkins, M. 81, 83
Wayne State University (WSU) 109–10, 115–16
websites 9, 13, 51
West 6, 8, 57, 67, 156

6 West Recording 103

Whitelaw, B. 17, 20, 25–30, 33–7, 40, 42, 153
Whitelaw, J. 42
Wider Opportunities for Women 107
Williamson, C. 81, 90, 103
Winter, C. 81
Wise, J. 137, 139
Woman Sound 91, 102
Women Airforce Service Pilots (WASP) 118, 147

Women in Aviation International/Pioneer Hall of Fame 119, 145
Women in Career Options 88–9, 94
Women in Technological History (WITH) 6
women-and-technology paradigm of empowerment 150–9
women's and gender studies 5, 13–14, 159
Women's Air Race 146
Women's Intellectual Association 47
women's movement 11, 57, 80–2, 89, 100–7, 109, 152
women's music 80–3, 88, 90, 93, 97, 101–6, 152, 156, 159
Women's National Sailing Championship 127–8
women's rights 45, 48, 50–2, 55, 64–5, 68–9, 71, 73, 75–6, 102, 158
women's studies 9, 88
Woodstock Peace and Music Festival 91
woodworking 42
workshops 98, 106, 153–4
World Championship Dragon Boat Festival 15
world music 83
Wright, T. 156

Xerox 124

Yemen 45
Young, C. 110
Your Concept Car (YCC) 8
YouTube 53

Zantop International Airlines 119
Zvereva, L. 117

For Product Safety Concerns and Information please contact our EU
representative GPSR@taylorandfrancis.com
Taylor & Francis Verlag GmbH, Kaufingerstraße 24, 80331 München, Germany

www.ingramcontent.com/pod-product-compliance
Lightning Source LLC
Chambersburg PA
CBHW070717220326
41598CB00024BA/3202